STRUCTURAL DESIGN OF CABLE-SUSPENDED ROOFS

ELLIS HORWOOD SERIES IN ENGINEERING SCIENCE

STRENGTH OF MATERIALS
J. M. ALEXANDER, University College of Swansea.

TECHNOLOGY OF ENGINEERING MANUFACTURE
J. M. ALEXANDER, R. C. BREWER, Imperial College of Science and Technology, University of London, J. R. CROOKALL, Cranfield Institute of Technology.

VIBRATION ANALYSIS AND CONTROL SYSTEM DYNAMICS
CHRISTOPHER BEARDS, Imperial College of Science and Technology, University of London.

COMPUTER AIDED DESIGN AND MANUFACTURE
C. B. BESANT, Imperial College of Science and Technology, University of London.

STRUCTURAL DESIGN AND SAFETY
D. I. BLOCKLEY, University of Bristol.

BASIC LUBRICATION THEORY 3rd Edition
ALASTAIR CAMERON, Imperial College of Science and Technology, University of London.

STRUCTURAL MODELLING AND OPTIMIZATION
D. G. CARMICHAEL, University of Western Australia.

ADVANCED MECHANICS OF MATERIALS 2nd Edition
Sir HUGH FORD, F.R.S., Imperial College of Science and Technology, University of London and J. M. ALEXANDER, University College of Swansea.

ELASTICITY AND PLASTICITY IN ENGINEERING
Sir HUGH FORD, F.R.S. and R. T. FENNER, Imperial College of Science and Technology, University of London.

INTRODUCTION TO LOADBEARING BRICKWORK
A. W. HENDRY, B. A. SINHA and S. R. DAVIES, University of Edinburgh.

ANALYSIS AND DESIGN OF CONNECTIONS BETWEEN STRUCTURAL JOINTS
M. HOLMES and L. H. MARTIN, University of Aston in Birmingham.

TECHNIQUES OF FINITE ELEMENTS
BRUCE M. IRONS, University of Calgary, and S. AHMAD, Bangladesh University of Engineering and Technology, Dacca.

FINITE ELEMENT PRIMER
BRUCE IRONS and N. SHRIVE, University of Calgary.

PROBABILITY FOR ENGINEERING DECISIONS: A Bayesian Approach
I. J. JORDAAN, University of Calgary.

STRUCTURAL DESIGN OF CABLE-SUSPENDED ROOFS
J. SZABÓ, Technical University of Budapest and L. KOLLÁR, Budapest City Council's Architectural and Town Planning Office.

CONTROL OF FLUID POWER, 2nd Edition
D. McCLOY, The Northern Ireland Polytechnic and H. R. MARTIN, University of Waterloo, Ontario, Canada.

TUNNELS: Planning, Design, Construction
T. M. MEGAW and JOHN BARTLETT, Mott, Hay and Anderson, International Consulting Engineers.

UNSTEADY FLUID FLOW
R. PARKER, University College, Swansea.

DYNAMICS OF MECHANICAL SYSTEMS 2nd Edition
J. M. PRENTIS, University of Cambridge.

ENERGY METHODS IN VIBRATION ANALYSIS
T. H. RICHARDS, University of Aston, Birmingham.

ENERGY METHODS IN STRESS ANALYSIS: With an Introduction to Finite Element Techniques
T. H. RICHARDS, University of Aston, Birmingham.

ROBOTICS AND TELECHIRICS
M. W. THRING, Queen Mary College, University of London.

STRESS ANALYSIS OF POLYMERS 2nd Edition
J. G. WILLIAMS, Imperial College of Science and Technology, University of London.

STRUCTURAL DESIGN OF CABLE-SUSPENDED ROOFS

J. SZABÓ
Professor of Mechanics
Technical University of Budapest

and

L. KOLLÁR
Head of Structural Engineering
Budapest City Council's Architectural
and Town Planning Office

Translation Editor:
M. N. PAVLOVIĆ
Leturer in Civil Engineering
Imperial College of Science and Technology
University of London

ELLIS HORWOOD LIMITED
Publishers · Chichester

Halsted Press: a division of
JOHN WILEY & SONS
New York · Brisbane · Chichester · Toronto

First published in 1984 by
ELLIS HORWOOD LIMITED
Market Cross House, Cooper Street, Chichester, West Sussex, PO19 1EB, England

and

AKADÉMIAI KIADÓ
Budapest, Hungary

The publisher's colophon is reproduced from James Gillison's drawing of the ancient Market Cross, Chichester.

Distributors:
Australia, New Zealand, South-east Asia:
Jacaranda-Wiley Ltd., Jacaranda Press,
JOHN WILEY & SONS INC.,
G.P.O. Box 859, Brisbane, Queensland 40001, Australia
Canada:
JOHN WILEY & SONS CANADA LIMITED
22 Worcester Road, Rexdale, Ontario, Canada
Europe, Africa:
JOHN WILEY & SONS LIMITED
Baffins Lane, Chichester, West Sussex, England
East European countries, Democratic People's Republic of Korea, People's Republic of Mongolia, Republic of Cuba and Socialist Republic of Vietnam
KULTURA, Hungarian Foreign Trading Company
P.O.B. 149, H—1389 Budapest, Hungary
North and South America and the rest of the world:
Halsted Press: a division of
JOHN WILEY & SONS
605 Third Avenue, New York, N.Y. 10016, U.S.A.

Translated by Lajos Bartha from the Hungarian *Függőtetők számítása* published by Műszaki Könyvkiadó, Budapest

British Library Cataloguing in Publication Data
Structural design of cable-suspended roofs. —
(Ellis Horwood series in engineering science)
1. Roofs, Suspension
I. Szabó, J. II. Kollár, L. III. Pavlović, M. N.
IV. Függőtetők számítása. *English*
721'.5 TH2417

Library of Congress Card No. 82-9299 AACR2

ISBN 0-85312-222-9 (Ellis Horwood Limited)
ISBN 0-470-27188-4 (Halsted Press)

Printed in Hungary

Contents

Introduction

The many articles and books which have been written on the subject of cable-suspended roofs tend to fall into two categories. The publications in the first category are concerned with imaginative conceptual design but lack a theoretical mechanical basis, whilst those of the second category are largely concerned with detailed discussions of general theories which designers generally find difficult to understand.

The present book is unusual in that it attempts to discuss the static and dynamic characteristics of suspended cable roofs in such a way that it will be readily intelligible to structural designers whilst maintaining sufficient generality of theoretical formulation. It is our firm intention to remove the subject's apparent mystique and if, as a result, we dispel the many common technical misconceptions, then we shall feel rewarded. Our clear aim is to render superfluous those design procedures which are excessively approximate. This we will do through a classification of the varying degrees of exactness in such a way that the engineer may adopt the analytical design procedure which is appropriate to the structure being considered and to the available computing facilities.

The types of cable networks, the problems associated with their statical analysis and the calculation procedures are surveyed in Chapter 1. In Chapter 2 a simple approximate method is presented for the calculation of single-layer doubly-curved cable nets and this method has its principal application to preliminary designs. In Chapter 3 the accuracy of this approximate method is examined through a numerical example. The exact method of calculation for suspended cable networks is described in Chapters 4 and 5. First the erected shape is determined and this is followed by the calculation of displacements and internal forces which result from load increments. A somewhat less exact method is then presented and this proves to be sufficiently accurate for most cases. Some additional design problems which are not generally presented in the technical literature are briefly discussed in Chapter 6. These are concerned with the stability of the edge ring, the vibration of the entire cable network and the problem of local flutter. The Appendix contains a brief summary of the necessary matrix algebra and also a discussion of the basic geometrical relations for bars the axes of which are curved in space. The latter is of value in the structural analysis of edge beams.

Many so far unpublished investigations are given in our book but we considered it impossible to give the full algorithm or computer program for every procedure. We gratefully acknowledge the enthusiastic support for the publication of this work provided by our associates through their valuable and time-consuming work. In particular we thank Miklós Berényi who performed the exact solution of the numerical example of Sections 3.7 and 6.2.4. and Example 5.5, and also Dr. Zsolt Gáspár, who solved Examples 4.6, 4.7 and 5.2. We have used the calculations of László Köröndi in the comparative numerical example of Chapter 3 (Sections 3.3 through 3.6) and in the numerical optimality analysis of Chapter 6 (Section 6.2) and again we express our thanks to him.

In developing the subject of this book, we strove for balance and it was not possible to explore in detail every type of calculation which is of practical interest. Examples of the latter are the problems of double layer nets and the problems which arise when cable structures are combined with relatively rigid elements such as beams. However the relationships presented in the book are of sufficient generality that, at least in principle, they make such calculations possible.

In the final formulation of this text we received extremely valuable assistance from the critical review of the manuscript by Dr. Elek Béres whose detailed comments related to almost every page. His work was almost that of a co-author; he greatly improved the intelligibility of the text and for this we express our grateful thanks. The book was written as a result of the close co-operation of the authors but Chapters 4 and 5 and the Appendix were the primary responsibility of János Szabó whilst Lajos Kollár was primarily responsible for Chapters 1, 2, 3 and 6.

In conclusion we should also like to express our gratitude to the Publishers, Ellis Horwood and Akadémiai Kiadó, the Publishing House of the Hungarian Academy of Sciences, for this fine publication.

1. General Discussion of Suspended Roofs

1.1. The Structural Design of Suspended Roofs

Structures without secondary load-bearing elements are frequently used for covering large spaces. From a statical point of view, this means that the structure is not made up of hierarchical parts (main girder, purlin, roofing) but is designed in such a manner that the main girder also serves as a space-enclosing (roofing) element. Two such types of structure are the *shell* and the *suspended roof.*
In its original form as a tent, the suspended roof solved both the load-bearing and the space-enclosing task by means of a single "canvas-like" structure. However, with modern suspended roofs (which are able to bridge large spans) the two functions are already separated: essentially, the supporting structure consists of a cable net in two directions, stretched on a rigid or flexible edge, while the roofing, whether it consists of panel-like elements or of a uniform "canvas", does not take part in load-bearing.
The structural action of a shell and that of a suspended roof can be contrasted as follows (refer to Fig. 1.1). An upper compression flange and a lower tension flange are necessary for bridging any span. In the *shell structure*

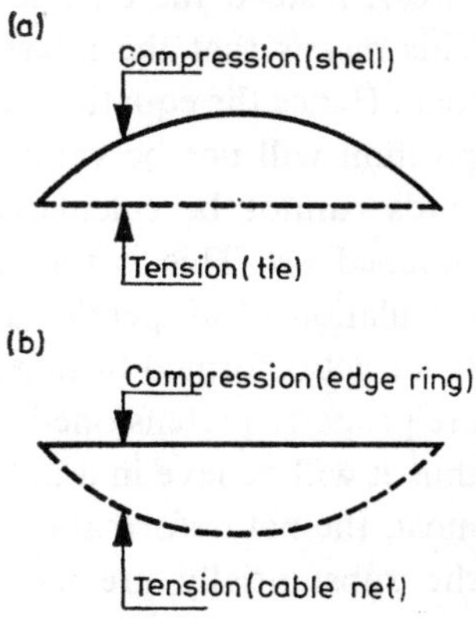

Fig. 1.1. Static behaviour of shells and suspended roofs

(Fig. 1.1a) the upper compression flange (the shell) extends to a large area, therefore it is thin and is susceptible to buckling, while the lower tension flang (the tie) is concentrated along the edge. On the other hand, in the *suspended roof* (Fig. 1.1b) the lower tension flange is the cable net itself; the upper compression flange is replaced by the edge ring or by the anchoring (in this system buckling is less likely to be a cause of concern).

The two structures differ also in that the *shell* is able to take *three* internal force components in its own plane, that is, it has a *rigidity* not only against two-dimensional tension-compression, but against shear as well. On the other hand, a *suspended roof* consisting of two cable rows can only take *two-dimensional tension;* the structure will take a small compression only when prestressed, but in no case will it take any shear, i.e. it has no shear rigidity. It follows from this that if the original form of the roof could equilibrate a given load only with the aid of shear, then the actual suspended roof is forced to change its shape in such a manner that in the new form it will be able to carry the load without shear. Clearly, the form adopted is one of the funicular surfaces of the load. Primarily, this explains why suspended roofs undergo large deformations. Thus, for example, the angle of the cables of the suspended roof of the stadium of the Munich Olympic Games changed by as much as 6° at some places compared to their position at the time of erection [16].

Another reason for the large deformation of suspended roofs is that the strength of the cable wires is very high, so that the corresponding elongations are considerable.

Finally, even relatively small movements of the edge may give rise to large net deformations. The flatter the cables, the larger the deformations become.

Due to the above reasons suspended roofs cannot be calculated on the basis of their original geometrical shape (as is usually the case for other structures); instead the change of geometry must be taken into consideration. This means that the internal forces do not vary linearly with the applied load. Hence the equations will also be non-linear, and the principle of superposition will not be valid, i.e. the stresses originating from the individual loads cannot be calculated independently from each other and then be summed up. This is the main reason for the unusual difficulties in the calculation of suspended roofs.

If the net is formed from a *three-dimensional system of cables* — and if these are properly pretensioned — the system will be rigid against shear also, and thus it will behave in a manner quite similar to that of membrane shells; at most, the net deformations will be greater than those of a shell because of the substantially greater permissible stresses in the cables. Hereinafter,

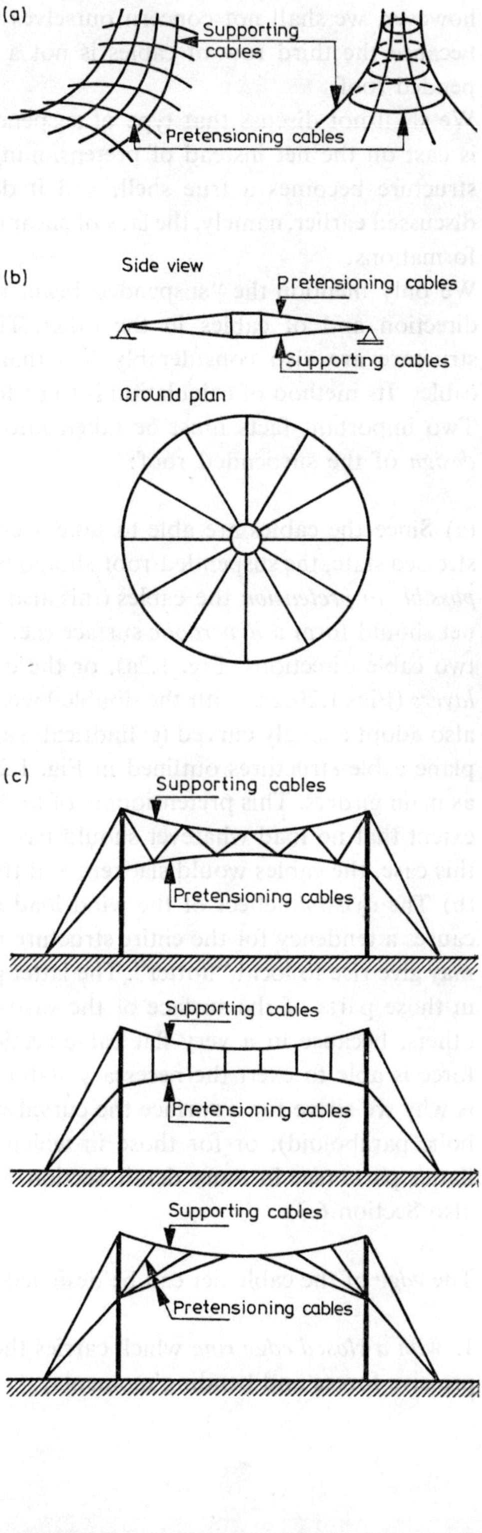

Fig. 1.2. Various types of suspended roofs

however, we shall not concern ourselves with this type of suspended roof because the third row of cables is not a characteristic feature of the suspended roof.

We shall not discuss that type of suspended roof either in which *concrete* is cast on the net instead of pretensioning the cables for, in this way, the structure becomes a true shell, and it does not show the characteristics discussed earlier, namely, the lack of shear rigidity and the resulting large deformations.

We only mention the "suspended beam roof" consisting of beams in one direction and of cables in the other. The deformations of this type of structure are also considerably less than those consisting exclusively of cables. Its method of calculation is to be found in [15] and [24].

Two important facts must be taken into consideration in the *geometrical design* of the suspended roof:

(a) Since the cables are able to take a compressive force only in the prestressed state, the suspended roof should be given such a shape that *it may be possible to pretension* the cables (this also without load). For this, either the net should form a *hyperbolic* surface (i.e. having opposite curvatures in the two cable directions, Fig. 1.2a), or the cables have to be arranged in *two layers* (Figs 1.2b, c). With the double-layer arrangement the cable system can also adopt a singly curved (cylindrical) surface in such a manner that the plane cable structures outlined in Fig. 1.2c are arranged beside each other as main girders. This pretensioning of the cables should be done to such an extent that no load whatever should terminate the tensile force because, in this case, the cables would slacken, and the net might "flutter" (see below).

(b) The *dynamic* effect of the wind load acting on suspended roofs partly causes a tendency for the entire structure to vibrate (Section 6.2) and partly may give rise to local "flutter". The latter phenomenon may occur primarily in those parts of the surface of the suspended roof which are flatter than others, because in a very flat cable section only a relatively great tensile force is able to exert the necessary stabilizing effect (see Section 6.3). This is why we strive for a surface the curvatures of which are constant (hyperbolic paraboloid), or for those in which greater cable forces arise in the flatter parts (see in more detail Section 1.2.2.3 on the erection shapes and also Section 6.3).

The *edge* of the cable net can be designed in three different ways:

1. *with a closed edge ring* which carries the cable forces by means of compressive forces and bending moments;

2. *with edge cables* which must be attached to a fixed support (to the ground) at some points, and which deform to a considerable extent under the various loads; and finally
3. *the ends of the cables* can be fixed to the ground *by trestles or by anchorings.*

Of these three edge solutions, the cable ends may be regarded as fixed in position only in the last case.
In general, the deformation of the edge ring cannot be neglected either, because it may have a considerable effect on the magnitude of the cable forces. This applies to the edge cables to an even greater extent.
We must concern ourselves also with the *location of the cables on the surface of the net.* The cables can be arranged on a doubly curved surface according to the following networks:

(a) a set of lines on the surface corresponding to a *perpendicular straight network in the ground plan projection;*
(b) the *principal curvature lines* of the surface: these constitute a (curvilinear) network on the surface so that the cables intersect at right angles;
(c) network *consisting of geodetic lines.* The geodetic lines are the "shortest" lines on the surface. If a yarn is stretched on the surface along the geodetic line, it does not slide laterally. Thus, the cables stretching on one another without friction endeavour to locate themselves along geodetic lines. If the cables are arranged according to this, then at the points of intersection only forces normal to the surface (and to each other) are transferred by them to each other. In general, the geodetic lines do not constitute a network intersecting at right angles either in ground plan projection or on the surface itself, and they are curves in the ground plan projection as well.
The difference between these three sets of lines can be dispensed with in the case of flat surfaces. On steeper surfaces, however, they constitute three distinctly different networks: thus it must be decided which network arrangement is to be adopted. The network corresponding to perpendicular straight lines in the ground plan makes the calculation technique easier (see Chapter 4); on the other hand, the cable net following lines of principal curvature has the advantage that it requires the smallest pretensioning forces against slackening (as the curvature of the cables is then greatest; cf. Section 6.3); another advantage is that at singular points (the peaks above the intermediate mast) the cables converge in a "natural" way.
(d) *From the point of view of erection technique,* the following type of cable net is advantageous: the net consists of a rectangular (square) network when developed into a plane; these squares become distorted when the

net is fitted onto the required surface. Such a system was used for the suspended roofs covering the stadium of the Munich Olympic Games [16].
What has been said so far about nets applies to curvilinear configurations drawn on the continuous (theoretical) suspended roof surface. Actual suspended roofs consist of cables lying at a finite distance from each other, so that only their joints can be on the theoretical surface; hence the cables themselves will be chord polygons of the curvilinear network.

1.2. Review of the System of Internal Forces and Calculation Methods of Cable Nets

1.2.1. Characteristics of the Static Behaviour of the Cables

In order to review the internal forces of cable nets and the various theories for their calculation, the characteristics of the static behaviour of such networks must be clearly understood. For the sake of simplicity, we begin by considering the response of a single cable.

1.2.1.1. The Large Deformation of a Cable under the Action of a Funicular Load

It is well known from rudimentary cable statics that a cable can assume only one kind of shape under a given load distribution; at most, the ordinates of the cable shape can be proportionately larger or smaller (depending on the horizontal component of the cable force). Hereinafter — irrespective of its intensity — the load distribution for which the funicular curve is the initial shape of the cable will be called the funicular load. If the intensity of a load having such a distribution is increased, this does not force the cable into "another shape", but may merely increase the ordinates of the sag.
The first characteristic of the behaviour of the cables is that *their deformation is relatively large,* and this affects the system of internal forces.
The calculation method which takes this effect also into consideration is called "second-order theory". The effect of the deformation on the internal forces consists of *a linear and a non-linear part,* which can be demonstrated by the following simple derivation:
Let us investigate the action of a cable, loaded by a uniform load q_0, the ends of which are fixed to two points at distance l from each other. Let the sag of the cable shape due to the load be f (Fig. 1.3); then the magnitude of the horizontal component of the cable force is

$$H_0 = \frac{q_0 l^2}{8f}. \tag{1.1}$$

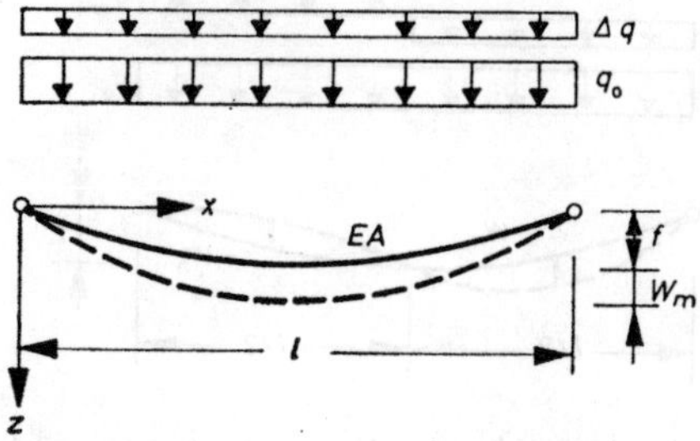

Fig. 1.3. Deflection of a cable due to uniform load

Let us increase the load by Δq, and let us see how the effect of the deformation on the internal forces manifests itself.

The cable elongates upon the effect of the excess load, and in its new position the sag increases by w_m, the horizontal cable force by ΔH. In accordance with this, (1.1) should be written in the following modified form:

$$H_0+\Delta H = \frac{(q_0+\Delta q)\,l^2}{8\,(f+w_m)}, \tag{1.2}$$

that is

$$(H_0 f+\Delta H f)+H_0 w_m \Delta H w_m = \frac{(q_0+\Delta q)\,l^2}{8}. \tag{1.3}$$

The two terms between brackets on the left side represent the cable force increment which can be calculated without the consideration of the deformation (i.e. on the basis of the original geometry), that is, according to the first-order theory. The next term expresses the effect of the change of the cable shape on the original cable force H_0: in consequence of the increase of the sag the H_0 arising from the original q_0 decreases because $q_0 l^2/8$ [Eq. (1.1)] must be divided by the increased sag $(f+w_m)$. This is the *linear* part of the second-order effect because in this the unknown quantity w_m occurs to the first power. Finally, the last term indicates that the cable force increment ΔH must also be calculated with the sag altered by w_m. This is the non-linear part of the second-order theory because the product of the two unknown increments appears here.

1.2.1.2. The Deformation of the Cable due to a Load other than the Funicular One

Especially large deformations occur if the load is not the funicular load of the initial form of the cable. Therefore, in such a case, the deformation of the cable cannot be neglected. Let us consider Fig. 1.4, which represents a flat cable loaded by a uniform load q_0, and on which an antisymmetric load q_{ant} is added. The force H_0 arising in the cable due to q_0 does not change upon the effect of the antisymmetric load q_{ant} (similarly as the force H_0 would

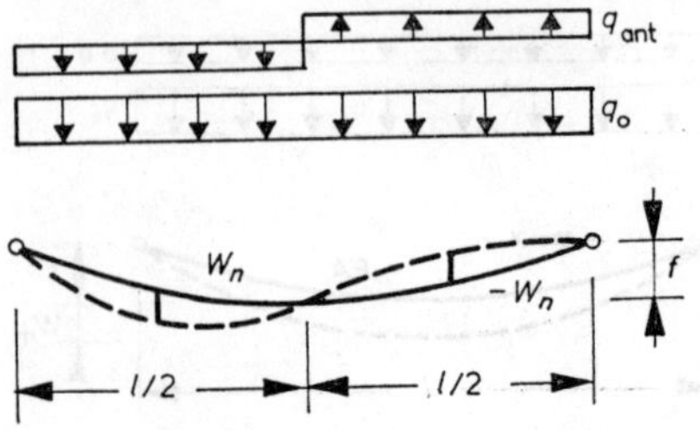

Fig. 1.4. Deflection of a cable due to antisymmetric load

also remain unchanged on an arch corresponding to the cable shape); thus the cable will be able to carry the load q_{ant} only by changing its shape. It will assume the shape indicated by the dashed line, the quarter-point deflection of which, w_n, is determined by the requirement [refer to Eq. (1.2)]

$$H_0 = \frac{q_{ant}(l/2)^2}{8w_n}. \tag{1.4}$$

It can be seen that now the change in shape of the cable cannot be neglected because the cable is able to carry the load q_{ant} only because of this change of shape.

It was assumed in the solution just described that the deflections w_n of the two quarter points are equal. This approximates reality only in the case of flat cables, while in the case of steep cables the two deflections differ noticeably. On the other hand, in the case of quite flat cables, these being nearly straight lines, the force H_0 can no longer remain constant because the cable is able to undergo only a limited deformation without elongation; hence the antisymmetric load can only be equilibrated with an unchanged force H_0. Thus, the approximate method described above reflects reality only in the case of "medium-flat" cables.

1.2.1.3. Horizontal Displacements of Cable Points

Horizontal displacement in the plane of the cable. The cable points are displaced in the plane of the cable not only vertically, but also horizontally. The horizontal displacement has two effects. On the one hand, the loads fixed to the cables (e.g. dead load) are also displaced horizontally, so that their intensity in the horizontal projection changes, and, in addition, their bending moment diagram is also altered. On the other hand, the cable joints are displaced relative to the bending moment diagram originating from the external load, and since it is this moment that determines the vertical position of the joints (i.e. the elevation of the joints) the shape of the cable net also changes. This effect of horizontal displacement has been known for a long time in the theory of arches [5, 20].

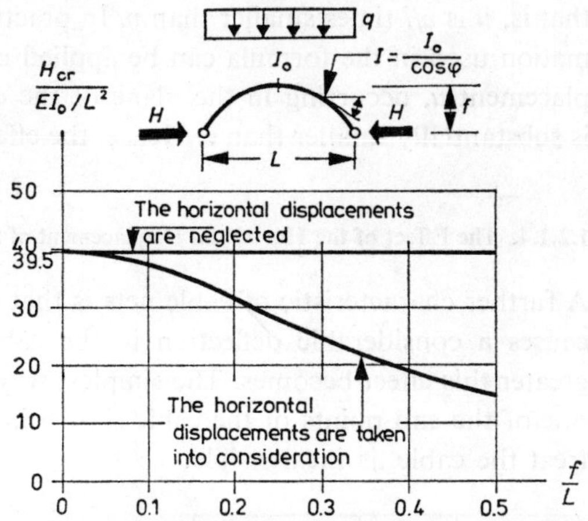

Fig. 1.5. Critical horizontal thrusts of an arch without and with consideration of horizontal displacements

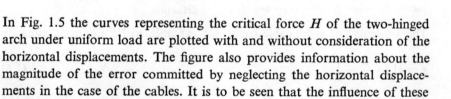

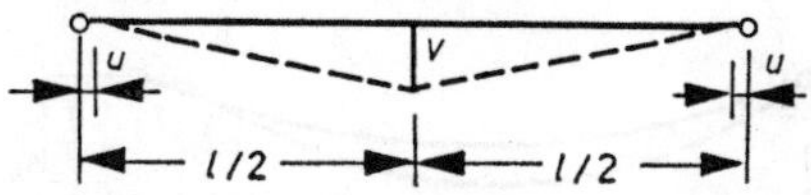

Fig. 1.6. Influence of a lateral displacement on those in the plane of the cable

In Fig. 1.5 the curves representing the critical force H of the two-hinged arch under uniform load are plotted with and without consideration of the horizontal displacements. The figure also provides information about the magnitude of the error committed by neglecting the horizontal displacements in the case of the cables. It is to be seen that the influence of these horizontal displacements, u, is small in the case of flat cables, but that it increases as the cable becomes steeper.

Horizontal displacement perpendicular to the plane of the cable. In the case of cables being in the vertical plane, the effect of a displacement v, perpendicular to this plane can be neglected. Namely, let us consider a cable originally straight in ground plan, the centre of which is displaced by an amount v in a direction perpendicular to the cable (Fig. 1.6). In the case of an inextensional cable, the displacement u of the end point in the direction of the cable can be expressed by the displacement v in the following manner:

$$u = \frac{l}{2} - \sqrt{\left(\frac{l}{2}\right)^2 - v^2} = \frac{l}{2}\left[1 - \sqrt{1 - \frac{v^2}{\left(\frac{l^2}{2}\right)}}\right] \approx$$

$$\approx \frac{l}{2}\left[1 - \left(1 - \frac{v^2}{2\left(\frac{l}{2}\right)^2}\right)\right] = \frac{v}{l}\,v, \tag{1.5}$$

that is, u is v/l times smaller than v. In practical cases $v/l \ll 1$ (the approximation used in the formula can be applied only in this case), thus the displacement u, occurring in the plane of the cable due to displacement v, is substantially smaller than v. Hence, the effect of v is negligible.

1.2.1.4. The Effect of the Horizontal Displacement of Supports

A further characteristic of cable nets is that the deformation of their edges causes a considerable deflection in the cables. The flatter the cable, the greater this effect becomes. The simplest way to illustrate this is to displace one of the end points of the cable shown in Fig. 1.7 inwards by Δl and to treat the cable as inextensible.

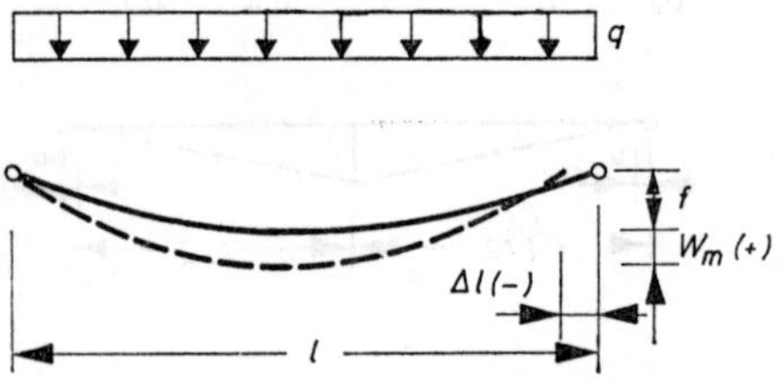

Fig. 1.7. Influence of a horizontal displacement of the support on the deflection

The arc-length of a *flat* cable (refer, for example, to [19]) is approximately

$$s = l\left(1+\frac{8}{3}\frac{f^2}{l^2}\right). \tag{1.6}$$

The support displacement Δl causes an increase w_m in the sag. The simplest way of relating these two quantities is to use the fact that the arc-length remains constant:

$$\frac{\partial s}{\partial f} w_m + \frac{\partial s}{\partial l}\Delta l = 0,$$

that is

$$\frac{16}{3}\frac{f}{l} w_m + \left(1-\frac{8}{3}\frac{f^2}{l^2}\right)\Delta l = 0$$

from which (since $f^2/l^2 \ll 1$),

$$w_m = -\frac{3}{16}\frac{l}{f}\left(1-\frac{8}{3}\frac{f^2}{l^2}\right)\Delta l \approx \frac{3}{16}\frac{l}{f}\Delta l. \tag{1.7}$$

It can be seen that in the case of small f/l ratios, the increase in sag w_m is substantially greater than the displacement Δl of the edge. Thus, this effect can be very significant.

1.2.2 Discussion of the Internal Force System of Suspended Roofs

1.2.2.1. Basic Assumptions

Before considering the structural action of suspended roofs and the different methods of calculation, the equations of suspended roofs (cable nets) consisting of a single-layer, bidirectional cable system will be briefly discussed. The following assumptions can be made for the sake of simplicity:

— the cables are "uniformly distributed" in the net, thus this constitutes a continuous, smooth surface;
— the supporting edge of the cables is rigid;
— the cables constitute a rectangular network in the ground plan projection;
— the horizontal displacement-components u and v of the cable points are negligible in comparison to vertical displacements w;
— only vertical loads Z act on the cable net. Also, since the cables are located in the vertical plane, the two cable rows transfer only vertical forces to each other. Thus, horizontal components H_x and H_y of the cable forces do not change along the individual cables; in the parallel cables, however, they can be different, i.e. $H_x(y)$ and $H_y(x)$, respectively.
For a knowledge of the static behaviour of the cable net, cable force components $H_x(y)$ and $H_y(x)$, as well as the deflection function $w(x, y)$ must be determined. For this, the following equations can be established (we indicate the differentiation with respect to x and y by a prime and a dot, respectively).

1.2.2.2. The Equation of Vertical Equilibrium

The equation of vertical equilibrium (similar to Eq. (1.3), but now stated as a differential equation for an arbitrary load and a bidirectional cable system) becomes:

$$(H_{0x}+\Delta H_x)(z+w)''+(H_{0y}+\Delta H_y)(z+w)^{\cdot\cdot}+Z=0. \qquad (1.8)$$

Here the starting (the so-called "erection") shape of the cable net is indicated by $z(x, y)$; this is the shape adopted when — besides the (negligible) self-weight — only pretensioning cable forces H_{0x} and H_{0y} act on the structure. The cable forces due to the load $Z(x, y)$ differ from these by ΔH_x and ΔH_y:

$$H_x = H_{x0}+\Delta H_x, \qquad (1.9a)$$

$$H_y = H_{y0}+\Delta H_y. \qquad (1.9b)$$

Equation (1.8) — similar to Eq. (1.3) — is non-linear because it contains the unknown cable force increments, ΔH, multiplied by the (also unknown) deflection function w.

Equation (1.8) in itself is evidently not sufficient for determining the unknown functions ΔH_x, ΔH_y and w. It is suitable, however, for obtaining the "erection shape" of the cable net. In this case we have to solve a linear equation (i.e. first-order theory).

1.2.2.3. Determination of the Erection Shape

Here the self-weight of the cables is neglected, so that only pretensioning forces H_0 act on the structure. Besides, $w=0$ by definition. Thus, by choosing z and H_0 in one direction in Eq. (1.8), the H_0 in the other direction is readily obtained from an algebraic equation. If, however, H_0 is fixed in both directions, then a differential equation of the elliptic type is obtained for z; this can also be solved by simple numerical methods (e.g. by relaxation: [8]), or, in many cases, its closed, analytical solution is available. In this manner it is possible to obtain a good estimate of the relationship between the erection shape of the cable net and the distribution of pretensioning forces H_0. In this case, the deformation of the edge girder is not included in the calculation, but it can be determined exactly after the erection shape of the net has been obtained.

Equation (1.8) determines the erection shape after substituting $w=0$, $Z=0$ as follows:

$$H_{0x}z''+H_{0y}z^{\cdot\cdot} = 0. \tag{1.10}$$

By adopting surfaces z in sum- (translational) and product-form a wide range of statically possible erection shapes can be investigated (refer to Schleyer [17]).

(a) In the case of z adopted as a *sum function:*

$$z(x, y) = g(x)+h(y). \tag{1.11}$$

By substituting this into (1.10), the following relationship is obtained:

$$\frac{H_{0x}(y)}{h^{\cdot\cdot}(y)} = -\frac{H_{0y}(x)}{g''(x)}, \tag{1.12}$$

which can be fulfilled only if both sides are equal to the same constant (C). Thus:

$$H_{0x}(y) = -Ch^{\cdot\cdot}(y), \tag{1.13a}$$

and

$$H_{0y}(x) = +Cg''(x). \tag{1.13b}$$

If, for example

$$z = \frac{k_x}{2}x^2 - \frac{k_y}{2}y^2, \tag{1.14}$$

where

$$k_x = \frac{8f_x}{l_x^2}, \tag{1.15a}$$

and

$$k_y = \frac{8f_x}{l_x^2}, \tag{1.15b}$$

the surface is a hyperbolic paraboloid, while k_x and k_y are the values of the curvatures valid at the origin. Then

$$H_{x0} = Ck_y, \tag{1.16a}$$

and

$$H_{y0} = Ck_x, \tag{1.16b}$$

that is, the pretensioning cable forces are constant in both directions, and their ratio depends on the ratio of the curvatures of the net in the x and y directions. It is characteristic for this hyperbolic paraboloid surface that it has curvatures (more exactly: second derivatives) and pretensioning cable forces of constant magnitude in each direction; thus, it is optimal from the point of view of eliminating flutter (see Section 6.3).

If the generatrices of the translation surface are hyperbolic cosine curves or circular arcs in the two directions, then the magnitude of the pretensioning cable forces increases from the centre towards the edges. If, however, the two generatrices have a cosine form, then the magnitudes of the pretensioning cable forces decrease as one moves away from the origin, and at the inflection point of the cosine curve they will become zero.

(b) In the case of z adopted in the form of a *product function:*

$$z(x, y) = g(x)h(y). \tag{1.17}$$

By substituting this expression into (1.10) and by separating the variables the pair of equations

$$g''(x) + \frac{H_{0y}(x)}{C}g(x) = 0, \tag{1.18a}$$

$$h^{\cdot\cdot}(y) - \frac{H_{0x}(y)}{C}h(y) = 0 \tag{1.18b}$$

is obtained, which gives the relationship between the pretensioning cable forces and the erection shape.

If, for example, constant pretensioning cable forces are adopted in both directions, we obtain

$$g(x) = a_1 \cos \alpha x + b_1 \sin \alpha x \tag{1.19a}$$

and

$$h(y) = a_2 \cosh \beta y + b_2 \sinh \beta y \tag{1.19b}$$

where a_1, a_2, b_1, b_2 are arbitrary constants, while

$$\alpha = \sqrt{\frac{H_{y0}}{C}}, \tag{1.20}$$

and

$$\beta = \sqrt{\frac{H_{x0}}{C}}. \tag{1.21}$$

From these, and using $\sin \alpha x = \cos\left(\frac{\pi}{2} + \alpha x\right)$, we obtain

$$z = a_1 \cos \alpha x \cosh \beta y \tag{1.22}$$

and

$$z = a_2 \cos \alpha x \sinh \beta y \tag{1.23}$$

which are surfaces, having a straight generatrix at $x = \pm\pi/2\alpha$ (and also at $y=0$ on the second surface) so that straight edges can be formed along these. Along these lines, however, the curvatures are zero in both directions, so that the surface is more susceptible to flutter than the hyperbolic paraboloid which possesses constant cable forces as well as constant curvatures.

In the case of pretensioning cable forces of varying intensity other surfaces are obtained. Thus, for example, the surface

$$z = C(a^2 - x^2)(b^2 + y^2) \tag{1.24}$$

also has straight edges along the $x = \pm a$ lines, but here (parallel to these edges) theoretically infinite pretensioning cable forces, H_{0y}, are obtained. This is advantageous from the point of view of flutter even though in reality only a cable force of finite magnitude acts on the net (cf. Section 6.3 on flutter).

In the case of an actual design problem,the erection shape should be selected by endeavouring to harmonize several differing criteria: namely, in addition to satisfying the functional requirements, it is desirable that small cable forces should arise under the loads (if possible), and, also, that the structure should be sufficiently safe against flutter. From among the erec-

tion shapes discussed so far, the sufficiently rigid ones against flutter can be selected on the basis of Section 6.3; the satisfaction of the other criteria, however, can be checked only after the calculation of the entire structure has been performed.
The determination of the erection shape of composite cable nets (e.g. supported by masts, or stretched on a more complex edge) is treated in [8] and in Chapter 4.

1.2.2.4. Compatibility Equations

Once the erection shape of the cable net has been obtained, then, in addition to the non-linear equation (1.8), the *compatibility* equations must also be satisfied in order to determine the deformation w taking place under arbitrary load Z, as well as the cable force increments ΔH_x, ΔH_y.
These equations express the fact that the values of the change in length of the cables calculated from the elongation (i.e. with the aid of Hooke's law) will be identical to those calculated from the deflections (i.e. geometrically). The change in length of the cables can be expressed in terms of the elongation as follows:
The elongation of an elementary cable arc-length $\mathrm{d}s_x$, with cross-section A_x, due to a cable force increment $\Delta S = \Delta H_x \dfrac{\mathrm{d}s_x}{\mathrm{d}x}$ and temperature change ΔT is:

$$\Delta \mathrm{d}s_x^{\mathrm{I}} = \frac{\Delta H_x}{EA_x}\frac{\mathrm{d}s_x}{\mathrm{d}x}\,\mathrm{d}s_x + \alpha_T \Delta T\,\mathrm{d}s_x, \tag{1.25}$$

where α_T is the coefficient of thermal expansion of the cable.
The elementary arc-length is related to the projected length $\mathrm{d}x$ according to the formula

$$(\mathrm{d}s_x)^2 = (\mathrm{d}x)^2 + z'^2(\mathrm{d}x)^2 \tag{1.26}$$

(this relationship holds for the original surface z).
For the elongation of the cable due to the deflection w, we note that

$$(\mathrm{d}s_x + \Delta \mathrm{d}s_x)^2 = (\mathrm{d}x)^2 + (z'+w')^2(\mathrm{d}x)^2. \tag{1.27}$$

Subtracting (1.26) from this:

$$2\mathrm{d}s_x \Delta \mathrm{d}s_x + (\Delta \mathrm{d}s_x)^2 = 2z'w'(\mathrm{d}x)^2 + w'^2(\mathrm{d}x)^2.$$

Since the inequality $\Delta \mathrm{d}s_x \ll \mathrm{d}s_x$ holds even in the case of large cable elongations, the second term of the left-hand side can be neglected in comparison with the first, and thereby we obtain the expression

$$\Delta ds_x^{II} = z'w' \frac{dx}{ds_x} dx + \frac{1}{2} w'^2 \frac{dx}{ds_x} dx. \tag{1.28}$$

The total elongation of the cables should be the same irrespective of whether we integrate (1.25) or (1.26) over the length of the cable. Hence, taking (1.26) also into consideration, we obtain:

$$\frac{\Delta H_x}{EA_x} \int_0^{l_x} (1+z'^2)\, dx + \alpha_T \Delta T \int_0^{l_x} \sqrt{1+z'^2}\, dx = \int_0^{l_x} \frac{z'w'}{\sqrt{1+z'^2}} dx$$

$$+\frac{1}{2} \int_0^{l} \frac{w'^2}{\sqrt{1+z'^2}} dx. \tag{1.29}$$

A similar relationship can be obtained for the cables in the y-direction. Thus, in this way — together with the non-linear equilibrium equation (1.8) — we have three differential and integral equations for the determination of the three unknown functions (ΔH_x, ΔH_y, w).

If the cable net is steep, horizontal displacements u and v (in the x- and y-directions respectively) of the cable points cannot be neglected; then, in addition to (1.8), the two equilibrium equations along the horizontal projections must also be written for the determination of the five unknown functions.

1.3. Survey of the Calculation Methods of Suspended Roofs

The different calculation methods differ from each other, partly in the type of approximation made, partly in the mathematical approach adopted.

1.3.1. Exact Methods

The most exact methods ([1], [7], [11], [18] as well as the method described in Chapters 4–5) take all the discussed effects into consideration (most of them directly, but the non-linear effect by iteration, in several steps). Essentially, the methods of the different authors vary in their approach to the iteration which takes the non-linear effect into consideration and also in the build-up of the calculation.

For the determination of the erection shape, the method discussed in Chapter 4 makes use of the advantageous property of nets rectangular in ground plan, namely that they enable the erection shape to be readily calculated. Finally, in Chapter 5, an iteration method will be presented for analyzing the effect of the loads; this approximates the correct result by relatively limited calculation work (depending on the accuracy required).

1.3.2. Approximate Methods

The various *approximate* calculations differ from one another primarily by what each neglects from among the previously discussed phenomena.

1.3.2.1. Approximate Methods for the Determination of the Distribution of Cable Forces

In order to simplify the resulting complex mathematical problem, it is advisable to neglect the horizontal displacements u and v. The calculation, having a limited exactness, is presented in Section 5.3 and is based on the following: it solves the problem leaving the horizontal displacements out of consideration, but it subsequently determines u and v from the compatibility equations, thus being able to obtain an estimate of the magnitude of the error committed.
If it is also assumed that the net is *flat*, then not only displacements u and v can be neglected, but the compatibility equation (1.29) also becomes simpler since $z'^2 \ll 1$, so that $1+z'^2 \approx 1$. However, the equilibrium equation (1.8) still remains non-linear (the equation contains cable forces ΔH multiplied by the unknown w). Owing to this non-linearity, the problem cannot, in general, be solved analytically. Eras and Elze [6] avoided this problem in such a manner [they neglected the second term of the right side of Eq. (1.29)] that they did not seek the deformation function w for a given load, but they adopted w in advance; then they determined analytically — assuming an infinitely rigid edge girder — first ΔH_x from the compatibility equation (1.29) (and ΔH_y from its symmetric equivalent) and then the load Z from (1.8). Each of these steps requires the solution of a linear problem. In this way, they were able to produce symmetric and antisymmetric loads for cases for which (in accordance with the non-linear theory) the exact deflection function and cable force system are already available.
The remaining approximate methods differ also in the method of solution, according to whether they regard the cable net as a continuum, or whether they perform the calculation by the discrete method (i.e. for every cable individually). Naturally, the cable net regarded as a continuum must also be solved (in most cases) by the discrete method (e.g. by finite differences),

but here we can take a coarser mesh than that corresponding to the joints of the cable net.

The basis of most solution methods is that they linearize Eq. (1.8), that is, they omit the terms containing the products of force increments ΔH and of the second derivatives of w, but they retain the products of the forces H_0 and the derivatives of w. In this way (i.e. retaining the linear effect of the deformation on the internal forces — see Section 1.2.1.1.), the effect of the non-funicular loads can also be taken into consideration (Section 1.2.1.2), and the neglected non-linear effect can then be followed by iteration. Schleyer's procedure belongs to this type of method [17]. In principle, it enables the taking into consideration of the deformation of the edges as well; however, in this paper he solves the problem only for an infinitely rigid edge girder. He regards the cable net as a continuum, but he solves the integro–differential equation by the discrete method (a finite difference equation system). The method is able to take into consideration not only vertical but horizontal loads as well.

Szmodits's method [30] is based on the same principles, but it assumes in advance that the edge girder is infinitely rigid. He also regards the cable net as a continuum when writing the non-linear integro–differential equation of deflection, which he transcribes into a difference equation system, and he follows non-linearity by iteration. He takes only vertical loads into consideration.

Bandel's procedure [2] differs from that of Szmodits in that he works with a discrete method: he writes down directly the difference equation system related to the joints.

1.3.2.2. Approximate Methods Adopting the Distribution of Cable Forces in Advance

Different cable forces may arise from the different loads in the individual cables of the net. This means that every cable force should be regarded as a separate unknown.

The calculation is greatly simplified if, besides the approximations listed so far, the distribution of the forces acting in the parallel cables is adopted in advance for each load type. Namely, in such a case it is no longer necessary to solve an equation system with many unknowns; instead, in most cases the calculation can be made extremely simple. Schleyer's approximate method [17] is based on this assumption. Now, the linear term due to the second-order effect, taken into consideration in his more exact theory, is also neglected. His method is valid only for a uniform load, but he gives a solution for the case when horizontal loads are present in addition to the vertical ones.

The approximate method described in [9] assumes the distribution of cable forces to be constant, and also omits the linear term due to the second-order effect. On the other hand, it takes the deformation of the edge into consideration. It is able to follow the entire second-order effect by iteration. It can be used not only for uniform, but for antisymmetric vertical loads too. This will be discussed in more detail in Chapter 2.
Roller [14] does not adopt the distribution of the cable forces as uniform but according to Fig. 1.8.

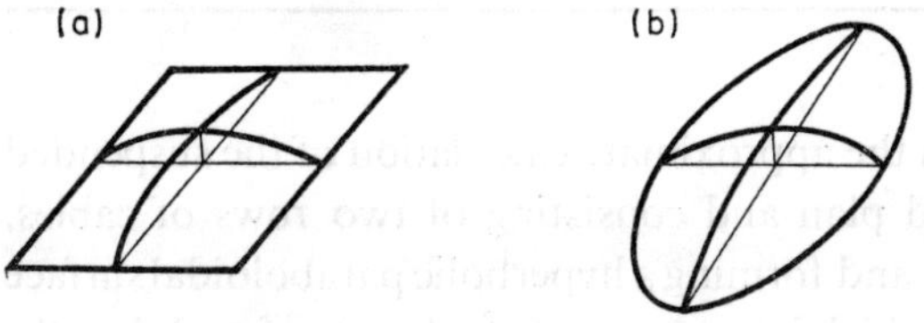

Fig. 1.8. Distribution of the cable forces assumed by Roller [14]

The error in the methods in which the distribution of the cable forces is chosen in advance will obviously be greater than that of the methods described in Section 1.3.2.1, where only the flatness of the cable net was assumed; the extent to which the latter assumption holds will also determine the magnitude of the error incurred. The comparative numerical example of Chapter 3 was worked out in order to illustrate this point. It presents a basis for determining the magnitude of the error committed, and shows that of the various possible cable force distributions it is the assumption of uniform distribution which approximates reality best (at least on the adopted hyperbolic paraboloid net, where the pretensioning cable forces required by the erection shape are also uniformly distributed).

2. Approximate Calculation of Suspended Roofs

This chapter is concerned with the approximate calculation of the suspended roof having an elliptic ground plan and consisting of two rows of cables, perpendicular in ground plan, and forming a hyperbolic paraboloidal surface [9]. The edge is a closed ring, which is supported by columns of such lengths that their rigidity against horizontal displacement can be neglected compared to the rigidity of the ring. This suspended roof has the simplest design on which it is possible to follow, by relatively simple calculations, the statical behaviour of suspended roof structures, and to illustrate all the characteristics of cable nets described in Section 1.2.1.

2.1. Approximate Assumptions

In order to make the calculation simpler, the following approximate assumptions are made:

(a) the cables are *flat*,

(b) *the loads are vertical*, and — at least in each quarter of the base area — they are uniformly distributed. (The actual distribution of the wind load can also be replaced by an equivalent uniform load).

(c) Of the bidirectional cable net only one representative cable *(or cable band of unit width)* is included in the calculation, assuming that all other cables parallel to them exhibit forces of the same magnitude;

(d) the internal forces and deformations of the *elliptic, spatial, curved edge ring* are calculated on the basis of a simple, *circular (planar) ring*. The compression of the edge ring is neglected, and only its bending deformation is taken into consideration;

(e) the *change of the geometry of the cable net* is not taken into consideration, that is, the internal forces are calculated on the *undeformed* structure. In this way, our equations will be *linear*, and the principle of superposition

will remain valid. The error committed due to this assumption can be reduced by calculating the deformation of the cable net and repeating the calculation by taking the changed shape as a basis for the iteration.
The last approximation, according to what was said in Section 1.2.1.2, can be applied only in the case of loads which do not cause shear forces in the original shape of the cable net, that is, those loads for which the undeformed net can be the funicular surface. Pretensioning and the uniformly distributed loads are examples of such loads and these will be discussed. In the case of antisymmetric loads it is always necessary to take into consideration the change of the cable shape, which in some cases (Fig. 2.1a and b) means only the change of shape of the row of cables in one direction, but in other cases (Fig. 2.1c) that of the rows of cables in both directions.

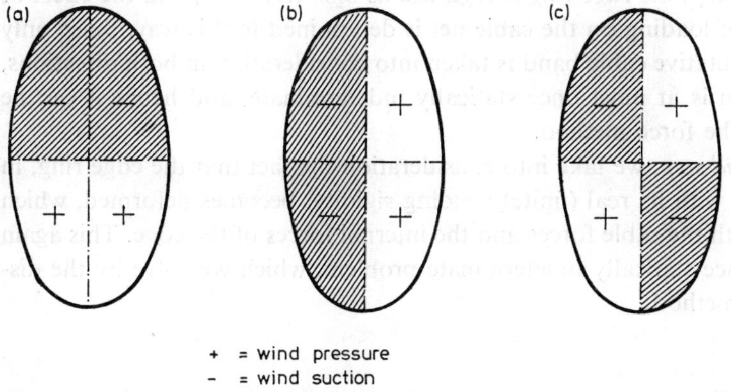

Fig. 2.1. Assumed distributions of wind pressure

2.2. Load Types

Actually, the following load types act on the cable net:

(a) dead weight (including the weight of the roofing),
(b) pretensioning,
(c) wind load,
(d) snow load,
(e) the (different) temperature change of the edge ring and the cable net.

Of these, the snow load and the dead weight (due to the flatness of the net) can be taken as uniformly distributed in ground plan. The wind load, regarded as acting vertically, in accordance with our assumptions, is approximated by antisymmetrically arranged, uniformly distributed partial loads. The calculation of the effect of the temperature difference can be made up of the solutions of the uniformly distributed load and the pretensioning. Thus, the cable net has to be solved for the following loading cases:

(a) uniformly distributed load,
(b) pretensioning,
(c) antisymmetric loads according to Figs 2.1a, b and c.

2.3. Method of Solution

The process of our calculation is as follows:
In the first step, the edge ring is regarded as infinitely rigid, and the effect of each type of loading on the cable net is determined in this way. Since only one representative cable band is taken into consideration in both directions, the problem is at most once statically indeterminate, and hence it can be solved by the force method.
In the second step, we take into consideration the fact that the edge ring, in accordance with its real (finite) bending rigidity, becomes deformed, which changes both the cable forces and the internal forces of the edge. This again means a once statically indeterminate problem, which we solve by the displacement method.

2.4. Detailed Treatment of the Approximate Calculation

We proceed as follows:
First, the statical properties of the structural elements of the suspended roof (i.e. the ring of finite rigidity and the cables), are outlined for a few basic cases. After this, the system of internal forces of the entire structure is "assembled" from these, and is solved for the different loading cases. Finally, the calculation method for the forces supporting the edge ring is discussed.

2.4.1. Analysis of the Edge Ring

Let us adopt the geometry of the suspended roof in accordance with Fig. 2.2. In this way, the equation of the surface will be as follows:

$$z = \frac{4f_x}{l_x^2} x^2 - \frac{4f_y}{l_y^2} y^2 \tag{2.1}$$

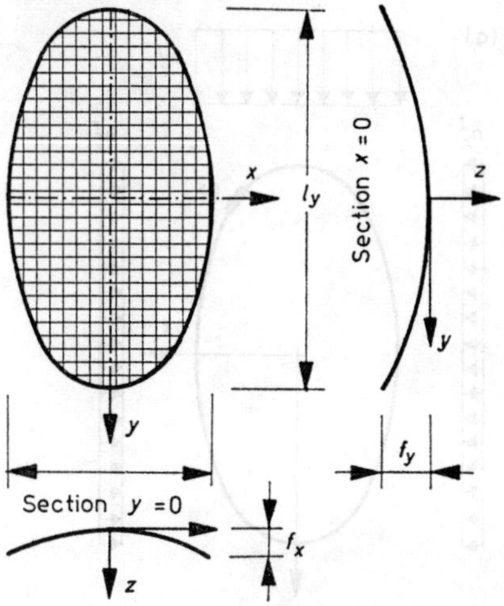

Fig. 2.2. The investigated type of cable roof

(hereinafter both f_x and f_y will always denote *positive* quantities). The equation of the horizontal projection of the edge is:

$$\frac{4}{l_x^2}x^2+\frac{4}{l_y^2}y^2=1. \tag{2.2}$$

As already mentioned, the spatial edge ring is approximated in the calculation by its projection on the horizontal plane. In the actual calculation we first determine the distribution of cable forces that does not cause bending in the ring but only compression; then we compute the cable force system which causes only bending of the ring. We need this separation because we neglect the compression of the ring, but take its bending deformation into consideration.

Thus, all force systems acting on the ring will be resolved into a part "producing pure compression" (marked with index I), which does not cause any bending of the ring, and another force system (marked with index II), "producing bending". As well as giving rise to bending in the horizontal plane, the latter force system also produces a small normal force of varying magnitude, but the integral of this normal force along the ring is zero.

First, it is shown that the horizontal projection of the elliptic edge is the funicular curve of all such cable force systems for which the x-and y-directional cable forces (n_x^{I} and n_y^{I}) are constant, and their ratio is:

$$\frac{n_x^{\mathrm{I}}}{n_y^{\mathrm{I}}}=\frac{l_x^2}{l_y^2} \tag{2.3}$$

(see Fig. 2.3a). n_x^{I} and n_y^{I} are regarded as positive if they denote *tension* in the cables, i.e. if they act *inwardly* on the ring in accordance with Fig. 2.3a.

In Fig. 2.3b, an elementary arc section of the elliptic edge is drawn. For this

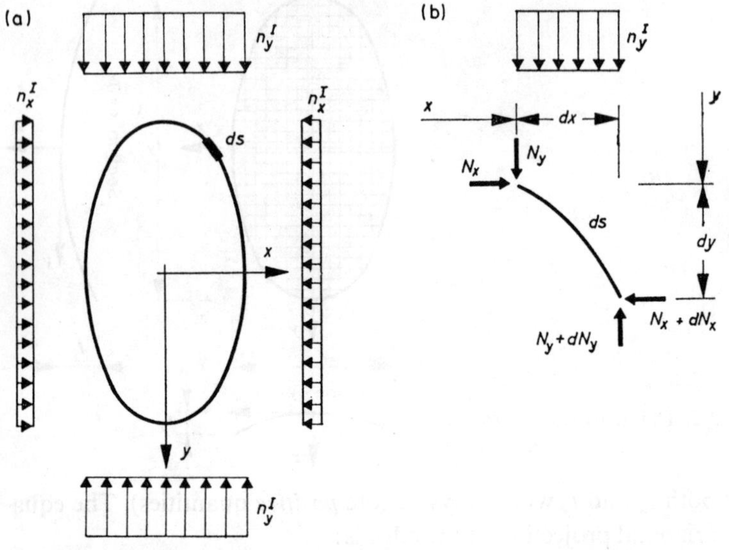

Fig. 2.3. Internal forces of the edge ring due to a "funicular" load system

it is possible to write down a geometrical equation expressing the fact that the direction of the normal force N always coincides with the tangent to the edge:

$$\frac{N_y}{N_x} = \frac{\mathrm{d}y}{\mathrm{d}x}. \tag{2.4}$$

Besides this, we have two equilibrium equations for the projections of forces in the x- and y-directions:

$$\mathrm{d}N_x = -n_x^{\mathrm{I}}\,\mathrm{d}y, \tag{2.5a}$$

and

$$\mathrm{d}N_y = n_y^{\mathrm{I}}\,\mathrm{d}x. \tag{2.5b}$$

Integrating these equations we obtain:

$$N_x = -n_x^{\mathrm{I}} y + N_{x0}, \tag{2.6a}$$

and

$$N_y = +n_y^{\mathrm{I}} x + N_{y0}. \tag{2.6b}$$

The integration constants N_{x0} and N_{y0} are equal to zero, since (Fig. 2.3a):

at $y = 0$: $N_x = 0$,

and

at $x = 0$: $N_y = 0$.

By substituting the expressions (2.6a, b) for N_x and N_y into (2.4), the differential equation of the funicular curve is obtained:

$$\frac{n_y^{\mathrm{I}} x}{-n_x^{\mathrm{I}} y} = \frac{\mathrm{d}y}{\mathrm{d}x}. \tag{2.7}$$

This is integrated to give:

$$n_y^{\mathrm{I}} \frac{x^2}{2} = -n_x^{\mathrm{I}} \frac{y^2}{2} + C. \tag{2.8}$$

Since at $y=0$, $x=\frac{l_x}{2}$, we obtain $C = n_y^{\mathrm{I}} \frac{l_x^2}{8}$.
If we divide both sides of (2.8) by C:

$$\frac{4}{l_x^2} x^2 + \frac{n_x^{\mathrm{I}}}{n_y^{\mathrm{I}}} \frac{4}{l_x^2} y^2 = 1. \tag{2.9}$$

And this, in the case when condition (2.3) is fulfilled, becomes the equation of the edge ellipse (2.2). Thus, our statement has been proved.
The edge ring carries the vertical force components as a continuous girder supported by the intermediate columns; thus its calculation follows standard procedures.
After these, in considering the deformation of the edge arc we need to obtain: that cable force distribution which causes only bending in the ring (in the

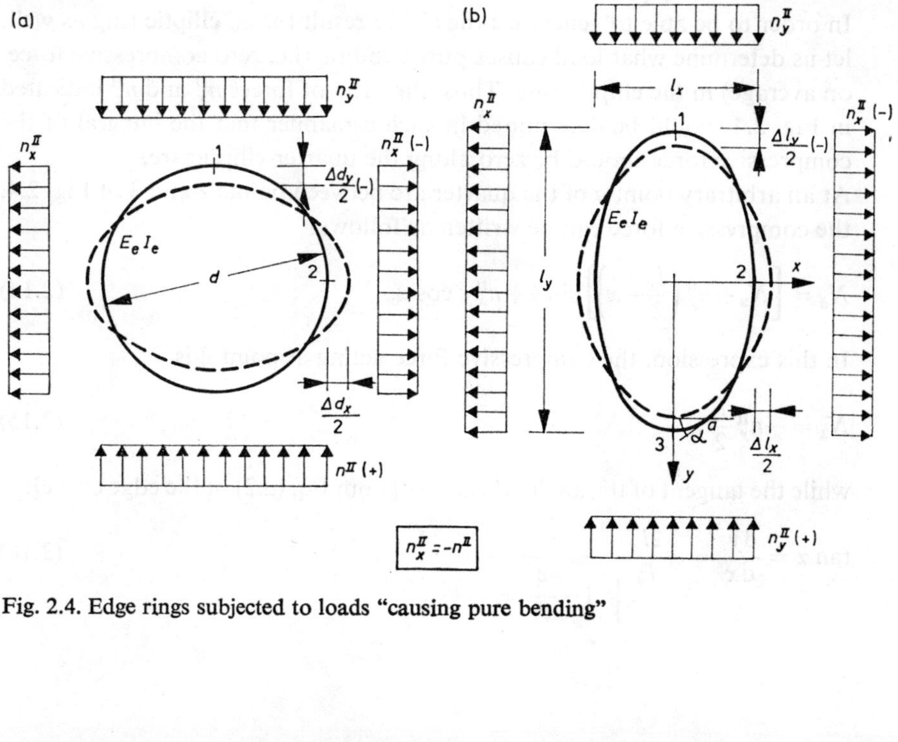

Fig. 2.4. Edge rings subjected to loads "causing pure bending"

horizontal plane), so that the compressive force in the ring is zero (on average).
For the sake of simplicity, the deformations of the elliptic ring are calculated on an "equivalent" circular ring with diameter d (which is the geometrical mean of the small and great axes of the ellipse):

$$d = \sqrt{l_x l_y}. \tag{2.10}$$

On a circular ring, pure bending is caused by the load corresponding to Fig. 2.3a, that is, if

$$n_x^{II} = -n_y^{II}. \tag{2.11}$$

If it is also assumed that the cross section of the edge ring is constant, then (e.g., in [13]) simple formulae can be found for the bending moments and deformations:

$$M_1 = -M_2 = \frac{n_y^{II} d^2}{8}. \tag{2.12}$$

(The bending moment is positive if it causes tension on the inside.) The changes in the diameters are:

$$\Delta d_x = -\Delta d_y = \frac{1}{48} \frac{n_y^{II} d^4}{E_e I_e} = M_1 \frac{d^2}{6 E_e I_e}. \tag{2.13}$$

Here, $E_e I_e$ is the bending rigidity of the edge ring (regarded as planar) in its own (horizontal) plane.
In order to be able to generalize the above result for an elliptic ring as well, let us determine what load causes pure bending (i.e. zero compressive force, on average) in the elliptic ring. Thus, the ratio of forces n_x^{II} and n_y^{II} indicated in Fig. 2.4 should be determined in such a manner that the integral of the compressive force should be zero along the quarter-elliptic arc.
At an arbitrary point a of the quarter arc between points 2 and 3 of Fig. 2.4, the compressive force can be written as follows:

$$N_a = \left[N_2 + n_y^{II}\left(\frac{l_x}{2} - x\right)\right] \sin\alpha + n_x^{II} y \cos\alpha. \tag{2.14}$$

In this expression, the compressive force acting at point 2 is

$$N_2 = -n_y^{II} \frac{l_x}{2}, \tag{2.15}$$

while the tangent of the angle α becomes [from Eq. (2.2) of the edge curve]:

$$\tan\alpha = \frac{dy}{dx} = -\frac{2 l_y}{l_x^2} \frac{x}{\sqrt{1 - \frac{4}{l_x^2} x^2}}. \tag{2.16}$$

The integral of the compressive force acting in the quarter arc between points 2 and 3 of the above figure is

$$\int_{(2)}^{(3)} N_a \, ds = 0, \tag{2.17}$$

so that, since $ds = \dfrac{dx}{\cos\alpha}$, and also using (2.14)–(2.15):

$$\int_{l_x/2}^{0} (-n_y^{\mathrm{II}} x \tan\alpha + n_x^{\mathrm{II}} y) \, dx =$$

$$= \int_{l_x/2}^{0} \left(\frac{2n_y^{\mathrm{II}} l_y}{l_x^2} \frac{x^2}{\sqrt{1 - \frac{4}{l_x^2} x^2}} + \frac{n_x^{\mathrm{II}} l_y}{2} \sqrt{1 - \frac{4}{l_x^2} x^2} \right) dx = 0. \tag{2.18}$$

Performing the integration

$$n_x^{\mathrm{II}} = -n_y^{\mathrm{II}} \tag{2.19}$$

is obtained, i.e. the same load ratio as in the case of the circular ring. Hereinafter it will be assumed that the bending moments according to (2.12) give the values (to a good approximation) for the elliptic ring as well. The change of diameter, however, cannot be considered to be the same in the two directions because in this case the arc-length of the ellipse cannot be assumed to remain constant.

For the determination of the ratio of the two diameter elongations (Δl_x and Δl_y; Fig. 2.4b), the approximate formula for the periphery of the ellipse is used. (Clearly, the nearer the ellipse approaches a circle, the better this approximation becomes — see, for example, [4]).

$$K_{\text{ellipse}} \approx \frac{\pi}{2} [1.5(l_x + l_y) - \sqrt{l_x l_y}]. \tag{2.20}$$

The invariance of the arc-length can be expressed as follows:

$$\frac{\partial K}{\partial l_x} \Delta l_x + \frac{\partial K}{\partial l_y} \Delta l_y = 0,$$

that is

$$\left(1.5 - \frac{l_y}{2\sqrt{l_x l_y}}\right) \Delta l_x + \left(1.5 - \frac{l_x}{2\sqrt{l_x l_y}}\right) \Delta l_y = 0.$$

From this

$$\Delta l_y = \frac{3 - \dfrac{l_y}{\sqrt{l_x l_y}}}{3 - \dfrac{l_x}{\sqrt{l_x l_y}}} \Delta l_x = -\alpha \Delta l_x, \tag{2.21a}$$

with the abbreviation

$$\alpha = \frac{3-\dfrac{l_y}{\sqrt{l_x l_y}}}{3-\dfrac{l_x}{\sqrt{l_x l_y}}}. \tag{2.21b}$$

Once the actual ellipse is specified (i.e. with l_x and l_y chosen), α becomes a constant.

Moreover, if it is assumed [as in (2.10)] that

$$\Delta d_x \Delta d_y = \Delta l_x \Delta l_y,$$

we obtain

$$\Delta l_x = +\frac{d_x}{\sqrt{\alpha}}, \tag{2.22a}$$

and

$$\Delta l_y = +\Delta d_y \sqrt{\alpha}. \tag{2.22b}$$

Hereinafter, the cable forces acting on the edge arc must be resolved in every case into a part causing pure compression (n_x^{I}, n_y^{I}) and a part causing pure bending (n_x^{II}, n_y^{II}). Since according to our assumption c (made in Section 2.1), cable forces n_x, n_y are always considered to be uniform in one direction, this decomposition can be simply carried out with the aid of the following equations (Fig. 2.5; as already mentioned the forces acting inward on the ring are positive):

$$n_x = n_x^{\mathrm{I}} + n_x^{\mathrm{II}}, \tag{2.23a}$$

$$n_y = n_y^{\mathrm{I}} + n_y^{\mathrm{II}}, \tag{2.23b}$$

$$\frac{n_x^{\mathrm{I}}}{n_y^{\mathrm{I}}} = \frac{l_x^2}{l_y^2}, \tag{2.23c}$$

$$n_x^{\mathrm{II}} = -n_y^{\mathrm{II}}. \tag{2.23d}$$

From these the part of the load causing compression only is:

$$n_x^{\mathrm{I}} = \frac{n_x + n_y}{1+\dfrac{l_y^2}{l_x^2}}, \tag{2.24a}$$

$$n_y^{\mathrm{I}} = \frac{n_x + n_y}{1+\dfrac{l_x^2}{l_y^2}}, \tag{2.24b}$$

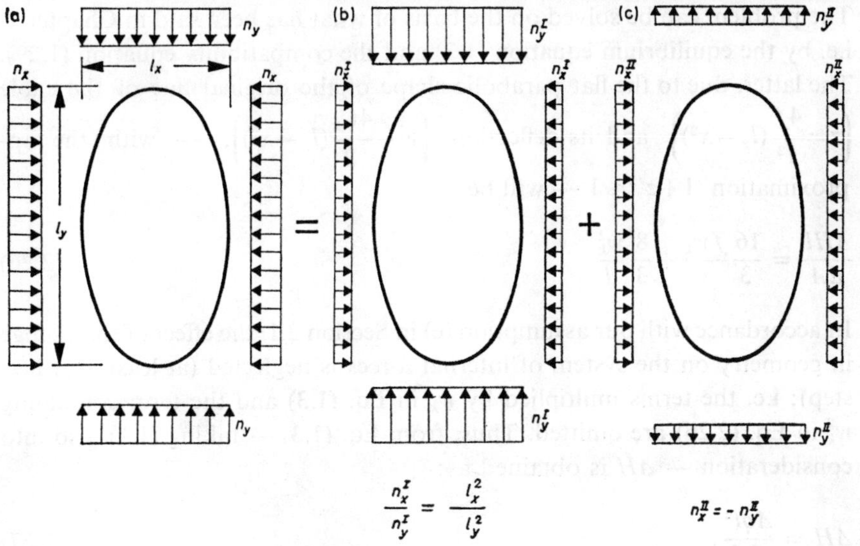

Fig. 2.5. Decomposition of the cable force system

and the part of the load causing bending:

$$n_x^{\text{II}} = \frac{n_x \dfrac{l_y^2}{l_x^2} - n_y}{1 + \dfrac{l_y^2}{l_x^2}}, \tag{2.25a}$$

$$n_y^{\text{II}} = -n_x^{\text{II}}. \tag{2.25b}$$

2.4.2. The Static Behaviour of the Cables

Let us examine the cable (or, alternatively, the 1 m wide cable band) acted upon by a uniform load q_0, as illustrated in Fig. 1.3. Let us fix the two ends of the cable at a distance l from each other, and at the same height. Under the load the cable adopts the flat shape defined by the sag f; we denote its (elongated) length by s. The cable force $n \equiv H$ (if acting on a unit width) is given by formula (1.1).

In order to obtain the statical behaviour of the cables, two basic problems must be solved.

First problem: What will be the change in the sag, the cable length and the cable force (w, Δs, $\Delta H = \Delta n$) when the load is increased by Δq (Fig. 1.3)?

The problem can be solved on the basis of what has been said in Chapter 1, i.e. by the equilibrium equation (1.3) and the compatibility equation (1.29). The latter, due to the flat parabolic shape of the original line of the cable $\left(z=\frac{4f}{l^2}(l_x-x^2)\right)$ and its deflection $\left(w-\frac{4w_k}{2}(l_x-x^2)\right)$, — with the approximation $1+z'^2 \approx 1$ — will be

$$\frac{\Delta Hl}{EA} = \frac{16}{3}\frac{fw_k}{l}+\frac{8}{3}\frac{w_k^2}{l}. \tag{2.26}$$

In accordance with our assumption (e) in Section 2.1, the effect of the change in geometry on the system of internal forces is neglected (at least as a first step); i.e. the terms multiplied by w_k in Eq. (1.3) and the term containing w_k^2 in Eq. (2.26) are omitted. Thus, from Eq. (1.3) — taking (1.1) also into consideration — ΔH is obtained as:

$$\Delta H = \frac{\Delta q l^2}{8f}. \tag{2.27}$$

Then, by substituting this result into Eq. (2.26) we obtain the increase in the sag:

$$w_k = \Delta H \frac{3}{16}\frac{l^2}{fEA_1} = \Delta q \frac{3}{128}\frac{l^4}{f^2 EA_1}, \tag{2.28}$$

where A_1 indicates the cross-sectional area of the cable band having unit width.

Since the cable is flat, the change Δs in the cable length is:

$$\Delta s = \frac{\Delta Hs}{EA_1}, \tag{2.29}$$

where the original cable length s is given by formula (1.6).

Second problem: If the two supports of the cable are brought nearer to each other by Δl (Fig. 1.6), what will ΔH, w_k and Δs be?

In accordance with our basic assumptions, we again neglect the effect of the change of geometry on the system of internal forces. Thus, the original cable force H_0 and the cable length s remain unchanged, and the change w_k of the sag is given by formula (1.7). (Δl is regarded as positive if it increases l, while w_k is regarded as positive if it is directed downwards, i.e. if it acts in the same direction as the axis $+z$.)

2.4.3. The Internal Forces of the Suspended Roof due to a Uniform Load

With the solution of the above two basic problems concerning the cables, we are now able to determine the internal forces due to the loads.
First the *uniform load* causing (approximately) constant cable forces will be discussed. The *displacement method* is used, that is, in the first step *the edge ring is regarded as being prevented from moving horizontally* and the cable forces are calculated in accordance with this. The flexibility of the ring is taken into consideration only in the second step.
During the calculation, the cables can be regarded as being able to take compression as well because the resulting compression will be eliminated by pretensioning.

2.4.3.1. Cable Forces Caused by the Uniform Load

The determination of the cable forces due to a *uniform, vertical load q* in the case of an *edge ring rigidly supported* against displacement is a statically indeterminate problem. The load q is distributed between the two cable rows:

$$q = q_x + q_y. \tag{2.30}$$

q_x and q_y are taken as being constant for each cable row because only the central cable bands are investigated. The ratio of the loads q_x and q_y can be established from the condition that the deflections of the two cable rows (or central cable bands, respectively) should be equal. Since the edge is regarded as rigid, formula (2.28) is used.

$$q_x \frac{3}{128} \frac{l_x^3 s_x}{f_x^2 (EA_1)_x} = q_y \frac{3}{128} \frac{l_y 3 s_y}{f_y^2 (EA_1)_y}. \tag{2.31}$$

From Eqs (2.30) and (2.31) q_x and q_y, and from these, through Eq. (1.1), n_{xq} and n_{yq} can be obtained.

2.4.3.2. The Deformation of the Edge Ring

The deformation of the edge ring is taken into consideration in the following manner:
If the imaginary support of the ring is removed, forces n_q deform the ring together with the cable net. As already mentioned, only the load part corresponding to Fig. 2.5c causes bending. This is the only deformation to be considered; the compression due to this loading component, as well as that due to the load component which causes pure compression (Fig. 25.b) is

neglected. Thus, we must first calculate the components of the cable forces n_{xq}, n_{yq} [these being determined by means of (2.30) and (2.31)] which cause bending only, i.e. n_{xq}^{II} and n_{yq}^{II} (2.25).
In accordance with the displacement method, the forces arising in the edge ring and the cable net due to the "unit" deformation $\Delta l_x = 1$ of the edge ring must be determined.
For a given Δl_x the change in diameter of the equivalent circular ring is, according to (2.21b) and (2.22),

$$d_x = -\sqrt{\alpha}\,\Delta l_x.$$

The system of forces acting on the ring, necessary to produce this deformation (see Fig. 2.4b) is, according to (2.13),

$$n_{yr}^{\mathrm{II}} = -n_{xr}^{\mathrm{II}} = \sqrt{\alpha}\,\Delta l_x \frac{48 E_e I_e}{d^4}, \tag{2.32}$$

where, in accordance with (2.10):

$$d = \sqrt{l_x l_y}.$$

Thereby the system of forces necessary for the deformation of the ring characterized by $\Delta l_x = 1$ is obtained.
The determination of the cable forces arising from the deformation Δl_x and [see (2.21a)]

$$\Delta l_y = -\alpha \Delta l_x$$

imposed on the cable net is also a statically indeterminate problem. Thus, let us first separate the two cable rows in order that they should be able to deform freely. In this way, the x-directional, central, 1 m wide cable band in the centre of the net exhibits a vertical deflection according to formula (1.7):

$$w_x^{\mathrm{I}} = +\frac{3}{16}\frac{l_x}{f_x}\Delta l_x, \tag{2.33a}$$

and the y-directional cable band undergoes a displacement

$$w_y^{\mathrm{I}} = +\frac{3}{16}\frac{l_y}{f_y}\alpha \Delta l_x. \tag{2.33b}$$

The cables that have moved away from each other must be fitted together again. For this, a vertical, internal force system, marked by p, uniformly distributed on the base area is introduced, in opposite sense to the two cable rows. (p is regarded as positive if it causes tension in both cable rows.) Due to these forces p, in the centre of the cables, according to (2.28), vertical displacements

$$w_x^{\mathrm{II}} = -p\frac{3}{128}\frac{l_x^3 s_x}{f_x^2(EA_1)_x}, \tag{2.34a}$$

and

$$w_y^{\mathrm{II}} = +p\frac{3}{128}\frac{l_y^3 s_y}{f_y^2(EA_1)_y} \tag{2.34b}$$

arise. The condition that the cables should fit together is as follows:

$$w_x^{\mathrm{I}} + w_x^{\mathrm{II}} = w_y^{\mathrm{I}} + w_y^{\mathrm{II}}. \tag{2.35}$$

Substituting expressions (2.33) and (2.34) into this, p is obtained as a function of Δl_x:

$$p = \frac{8\Delta l_x\left[+\dfrac{l_x}{f_x} - \dfrac{l_y}{f_y}\alpha\right]}{\left[\dfrac{l_x^3 s_x}{f_x^2(EA_1)_x} + \dfrac{l_y^3 s_y}{f_y^2(EA_1)_y}\right]}. \tag{2.36}$$

Then, from this, on the basis of (1.1), cable forces

$$n_{xc} = \frac{pl_x^2}{8f_x}, \tag{2.37a}$$

and

$$n_{yc} = \frac{pl_y^2}{8f_y} \tag{2.37b}$$

are obtained. The cable forces (2.37) are again resolved into components causing compression and bending in the ring respectively, by means of formulae (2.24) and (2.25). Thus n_{xc}^{II} and n_{yc}^{II}, related to the bending of the ring, are obtained (as functions of Δl_x).

Now, according to the displacement method, the entire structure must be made to undergo a deformation Δl_x of such magnitude that the forces n_{xr}^{II} and n_{xc}^{II} (and n_{yr}^{II} and n_{yc}^{II}, respectively) required for this should be equal to n_{xq}^{II} (and n_{yq}^{II}, respectively) originating from the load:

$$n_{xq}^{\mathrm{II}} = n_{xr}^{\mathrm{II}} - n_{xc}^{\mathrm{II}}. \tag{2.38a}$$

n_{xc}^{II} must be taken here with a negative sign because this is the force acting on the *cables*, which is the opposite of force n_{xr}^{II} acting on the *ring*. [Because of (2.11), it is superfluous to write the same equation for the forces in the y-direction]. Δl_x is calculated from Eq. (2.38a); then — with the aid of (2.32) and (2.36)–(2.37) — the corresponding ring and cable forces, respectively, are obtained. The forces n_{xq}^{II} and n_{yq}^{II} calculated originally decrease, of course, owing to the deformation of the ring, and the final cable forces will be equal to the forces acting on the ring. Thus, the magnitude of the final cable forces (causing bending) is

$$n^{\mathrm{II}}_{xc\,\mathrm{final}} = n^{\mathrm{II}}_{xq} + n^{\mathrm{II}}_{xc}. \tag{2.38b}$$

2.4.4. Internal Forces due to Pretensioning

In the case of pretensioning, the magnitude of the tensile force is given. Let us pretension, for example, the x-directional cables with a force n_{xP}, which is constant for all the cables running in the x-direction. In this case, according to (1.1), vertical forces

$$p = n_{xP}\frac{8f_x}{l_x^2} \tag{2.39}$$

are transferred to the y-directional cables, so that, in these, cable forces

$$n_{yp} = \frac{pl_y^2}{8f_y} = n_{xP}\frac{f_x}{f_y}\frac{l_y^2}{l_x^2} \tag{2.40}$$

arise. Since the cable forces (2.40) were obtained exclusively with the aid of the equilibrium equation (1.1), n_{yP}, as well as *all cable forces and internal forces, are independent of the rigidity (and hence the deformation) of the ring.* Thus, the problem is statically determinate.

In order to determine the deformation of the ring, the part of the load causing bending is calculated from n_{xP} and n_{yP} with the aid of expressions (2.25); then the deformation of the ring follows from formula (2.13).

2.4.5. The Effect of a Temperature Change

The different temperatures of the edge ring and the cable net cause internal forces in the structure. The effect is quite similar to that of pretensioning; namely, if the net is imagined as being separated from the ring, then, as a consequence of the temperature difference, the two elements will move away from each other uniformly along the circumference, and the ring will be larger than the perimeter of the net. Thus, the net must be pretensioned in order that its edge should be made to fit again to the ring. This is achieved by the following steps:

First, the two cable rows are separated; then, displacements are imposed on them so that they should fit to the edge ring. During this, the cables undergo sag changes w_x^I and w_y^I, which can be calculated from formula (1.7). In order to eliminate this separation, a uniformly distributed system of internal forces p must be introduced which — taking the ring as rigidly supported for the time being — can be obtained from Eq. (2.36). The deformation of the ring can be taken into consideration exactly in the manner detailed in Section 2.4.3.

2.4.6. The Analysis under Antisymmetric Loads

Of the load distributions shown in Fig. 2.1, cases a and b can be treated in the same manner; thus only two types of load distribution should be investigated, as shown in Fig. 2.6a and b. The loads of intensity q (uniform in each of the shaded sections) are, again, taken as being vertical.

In investigating the effect of the uniform load and pretensioning in Fig. 2.6 we could start from the unchanged geometry of the net because this is a funicular surface for the uniform load. In its original shape, however, the net is unable to balance an antisymmetric load; thus it must assume a new shape, namely that of the funicular surface for the antisymmetric load so as to be able to balance the load by means of bidirectional cable forces only, i.e. without shear.

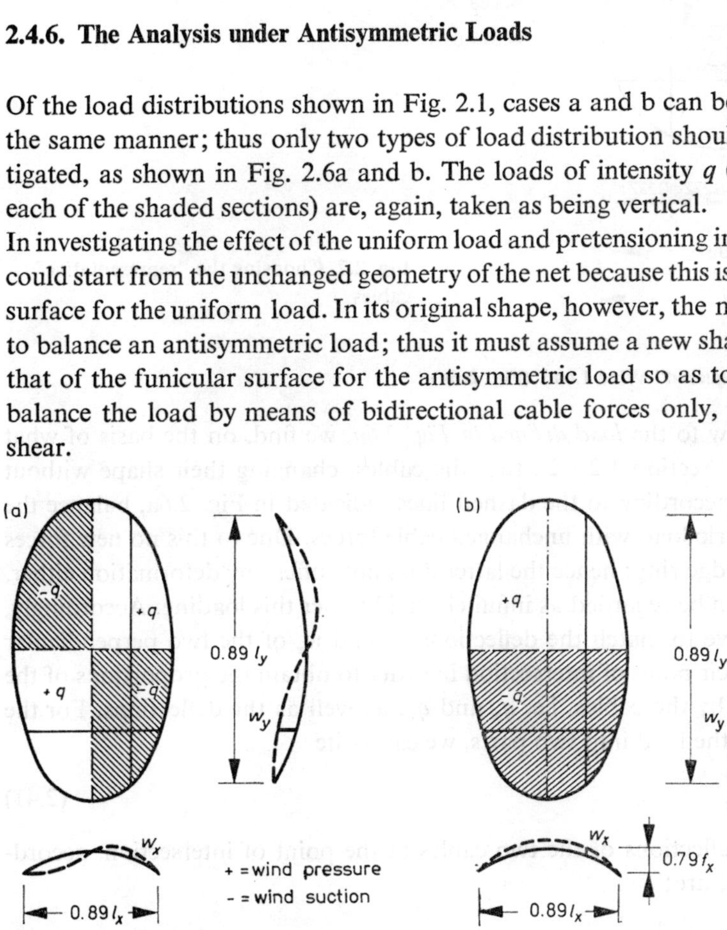

Fig. 2.6. Influence of antisymmetric wind loads

First of all, two "characteristic" cables (each consisting of a 1 m wide cable band) should be selected in both directions, and these are taken as a basis for the calculation. (It is not possible to start now from the middle cables because no forces arise in them). It is convenient to take as a basis those cables which intersect each other *at their quarter points*. In the case of a circular ground plan, this takes place with the ratios indicated in Fig. 2.7. Due to the affinity, in the case of an elliptic ground plan, the same ratio can be retained (see Figs 2.6a and b).

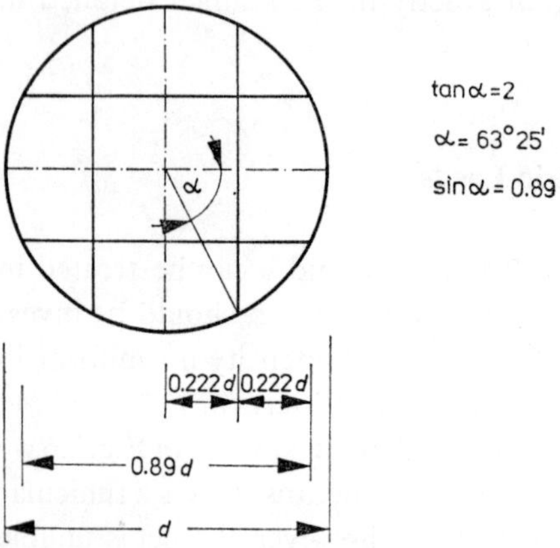

Fig. 2.7. Choosing the "representative" cables

2.4.6.1. "Chequered" Wind Load (Fig. 2.6a)

Coming now to the *load defined in Fig. 2.6a,* we find, on the basis of what was said in Section 1.2.1.2., that the cables, changing their shape without elongation according to the dashed lines indicated in Fig. 2.6a, balance the antisymmetric load with unchanged cable forces. Due to this no new forces act on the edge ring; hence the latter does not suffer any deformation either, and so it can be regarded as infinitely rigid under this loading. Accordingly, we only have to match the deflections w_x and w_y of the two perpendicular cables at their point of intersection in order to obtain the proportions of the load taken by the cables, i.e. q_x and q_y, as well as the deflections. For the division of the load into two parts, we can write

$$q_x + q_y = q, \tag{2.41}$$

while the deflections of the two cables at the point of intersection, according to (1.1), are:

$$w_x = \frac{q_x \left(\frac{0.89 l_x}{2}\right)^2}{8 n_x}, \tag{2.42}$$

$$w_y = \frac{q_y \left(\frac{0.89 l_y}{2}\right)^2}{8 n_y}. \tag{2.43}$$

Here, n_x and n_y denote the cable forces arising from the previous loads. Making the two deflections equal and making use of (2.41), it is a simple matter to obtain q_x and q_y, and from these the deflection of the point of intersection follows.
Thereby the deformation of the cable net is solved.

2.4.6.2. Antisymmetric Wind Load (Fig. 2.6b)

The solution of the case shown in Fig. 2.6b is somewhat more complicated because only the y-directional cable is able to change its shape without elongation in the manner required by the load, as the x-directional cable is forced to elongate. Taking the edge ring as infinitely rigid (for the time being), the deflections of the separated cables are calculated, and then, fitting these together, the load distribution, the change in the x-directional cable force, as well as the common deflection are determined.
The deflection of the y-directional cable at the quarter point is again given by formula (2.43); this cable force does not vary, since the cable carries the load by virtue of the change in its shape only. On the other hand, the deflection of the x-directional cable can be determined by the approximate formula (2.28). At its quarter point, 3/4 of the greatest deflection occurs:

$$w_x = \frac{3}{4} q_x \frac{3}{128} \frac{(0.89 l_x)^4}{(0.79 f_x)^2 (EA_1)_x} = \frac{9}{512} q_x \frac{l_x^4}{f_x^2 (EA_1)_x}. \tag{2.44}$$

Here we make the approximation that the arc-length of the representative cable is 0.89 times the arc-length of the longest cable.
By making the two deflections equal, and using relationship (2.41), the proportions of the load acting on the individual cables, the common deflection and the force increment that arises in the x-directional cable are obtained in a simple manner [the force increment is computed on the basis of the approximate relationship (2.27)].
The force increment arising in the x-directional cable row transmits the load shown in Fig. 2.8 on the ring [if it is assumed that not only the quarter point cable band, but every cable carries the internal force calculated above, and that the elliptic ring is substituted for by the circular ring according to relationship (2.10)]. According to [13], the internal forces of the ring are as follows ($+M$ causes tension on the inside, and $+N$ implies tension in the ring):

$$M_1 = -M_3 = +0.19 n_x d^2, \tag{2.45a}$$

$$M_2 = M_4 = 0, \tag{2.45b}$$

$$N_1 = -N_3 = 0.288 n_x d, \tag{2.45c}$$

$$N_2 = N_4 = 0. \tag{2.45d}$$

It is obvious that the ring also changes its shape in this case but this change is smaller than in the case corresponding to Fig. 2.4b. The shape of the deformation is shown by the dashed line of Fig. 2.8. This increases the deflection of the x-directional cables; thus it relieves them to the detriment of the y-directional cables. The latter, on the other hand, carry the load q_y acting on them without a force increment; but only by deformation; therefore, the taking of q_y as being smaller than in reality (because of the assumption of the infinitely rigid edge ring) does not cause a reduction in the safety of the structure. From the point of view of the x-directional cables, however, we approximated on the safe side.

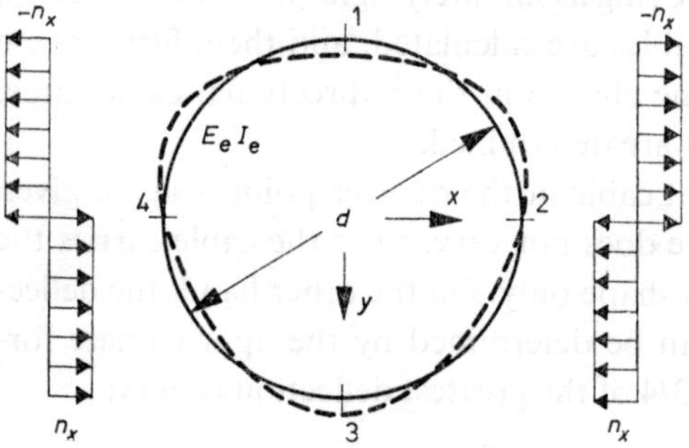

Fig. 2.8. Deformation of the edge ring due to antisymmetric wind load

The error is reduced by the fact that formula (2.44), gives the deflection on the basis of the original geometry, providing a deflection greater than the real one.

There is nothing to prevent the taking into consideration of the effect of the deformation of the edge in the manner discussed in the case of the uniform load. However, we will not discuss this here in detail.

2.4.7. Determination of the Forces Transferred to the Supports

The vertical forces *acting on the supports* (to be referred to from now on simply as support forces) arise partly from the vertical component of the cable forces, partly from the compressive force acting in the edge ring (the latter exerts a vertical resultant on the supports owing to the vertical curvature of the ring).

To determine the vertical component of the cable forces we need the vertical slope of the cables at their point of contact with the edge ring, that is, the tangent of the surface parallel to the two cable rows must be determined. Differentiating Eq. (2.1) of the surface with respect to x and y, the following values for the tangent are obtained:

$$\frac{\partial z}{\partial x} \equiv z' = \frac{8f_x}{l_x^2}\,x, \tag{2.46a}$$

$$\frac{\partial z}{\partial y} \equiv z' = -\frac{8f_y}{l_y^2}\,y. \tag{2.46b}$$

Thus, the first component v^{I} of the vertical support force v acting on a differential element of arc-length ds along the perimeter can be obtained in the following manner:

$$v^{\mathrm{I}}\,\mathrm{d}s = -n_x z'\,\mathrm{d}y - n_y z^{\cdot}\,\mathrm{d}x,$$

that is, for a unit arc-length:

$$v^{\mathrm{I}} = -n_x z'\frac{\mathrm{d}y}{\mathrm{d}s} - n_y z^{\cdot}\frac{\mathrm{d}x}{\mathrm{d}s}. \tag{2.47}$$

The signs are: v is *positive* if it acts, from the ring, downwards on the supports; and the perimeter s must be measured clockwise (Fig. 2.9).

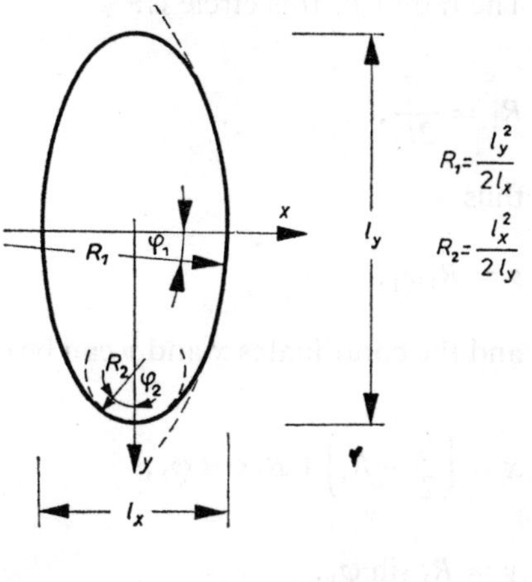

Fig. 2.9. Determination of the radii of the osculating circles

For the second component of the support force, v^{II}, we need the second derivative of the centre line of the edge ring in the vertical plane. This is evidently the same as the second derivative of the surface z taken with respect to the edge curve (of course, the horizontal projection of the arc length is implied here): $\frac{\mathrm{d}^2 z}{\mathrm{d}s^2}$. This must be multiplied by the tensile (or compressive) force acting in the edge ring in order to obtain the support force v^{II} acting on the unit arc-length:

$$v^{II} = N\frac{d^2z}{ds^2} \tag{2.48}$$

($+N$ means tension).

Since the formulae for both v^I and v^{II} include the arc-length, in the case of an elliptic perimeter it is not possible to write the support force expression v in a closed form even in the simplest case, i.e. when the cable forces are uniformly distributed. It is possible, however, to determine the values at the end points of the two principal axes of the ellipse:

At the end point of the small axis $\left(x=\frac{l_x}{2};\ y=0\right)$:

$$\frac{dx}{ds} = 0, \tag{2.49a}$$

$$\frac{dy}{ds} = +1. \tag{2.49b}$$

For the determination of $\frac{d^2z}{ds^2}$, the ellipse must be replaced by its osculatory circle in the vicinity of the point, according to Fig. 2.9.

The radius of this circle is:

$$R_1 = \frac{l_y^2}{2l_x}, \tag{2.49c}$$

thus

$$s = R_1\varphi_1, \tag{2.49d}$$

and the coordinates x and y can be expressed by φ_1 in the following manner:

$$x = \left(\frac{l_x}{2} - R_1\right) + R_1 \cos\varphi_1, \tag{2.49e}$$

$$y = R_1 \sin\varphi_1. \tag{2.49f}$$

By substituting (2.49d–f) into (2.1) and differentiating twice with respect to $s=R_1\varphi_1$, the following expression is obtained:

$$\frac{d^2z}{R_1^2\, d\varphi_1^2} = \left(-\frac{4f_x}{l_x^2} - \frac{4f_y}{l_y^2}\right) 2\cos 2\varphi_1 - \frac{4f_x}{l_x^2}\,\frac{\frac{l_x}{2} - R_1}{R_1}\, 2\cos\varphi_1,$$

and in the case of $s=\varphi_1=0$, substituting the expression of R_1 according to (2.49c):

$$\left(\frac{d^2z}{ds^2}\right)_{\substack{x=\frac{l_x}{2}\\ y=0}} = -8\frac{f_x+f_y}{l_y^2}. \tag{2.49g}$$

Incidentally, this result can also be obtained by differentiating the projection of the edge curve on the plane yz with respect to y.

At the end point of the great axis, on the other hand, $\left(x=0;\quad y=\frac{l_y}{2}\right)$:

$$\frac{dx}{ds} = -1, \tag{2.50a}$$

$$\frac{dy}{ds} = 0. \tag{2.50b}$$

The radius of the osculatory circle is:

$$R_2 = \frac{l_x^2}{2l_y}, \tag{2.50c}$$

and, as above, the second derivative taken with respect to the arc-length is:

$$\left(\frac{d^2z}{ds^2}\right)_{\substack{x=0\\ y=\frac{l_y}{2}}} = 8\frac{f_x+f_y}{l_x^2}. \tag{2.50d}$$

By knowing the values of n_x and n_y at these two points under the various loads, it is possible to determine the entire support force v at the two apices. It should be pointed out that — contrary to expectations — the greatest support force acting downwards is obtained at the deepest point ($\varphi=\pi/2$) even though here the cables themselves pull the ring upward. This shows that the support force arising from the force N acting in the ring (due to the vertical curvature) is greater than the direct support force of the cables, and thus the sign is determined by the curvature of the ring. Consequently, the support force will vary along the perimeter of the ring according to the curve shown in Fig. 2.10.

This distribution of the support forces makes it possible to adopt the following interesting structural design. Since the support force v raises the highest point of the suspended roof upwards, it is possible *not to support* the two sections of the edge protruding higher, but to extend them as "cantilevers" from the deeper, supported part (Fig. 2.11). Thereby the structure can be imparted a light appearance. Also, the statical behaviour can simply be formulated in such a manner that the two protruding roof parts are cantilevers

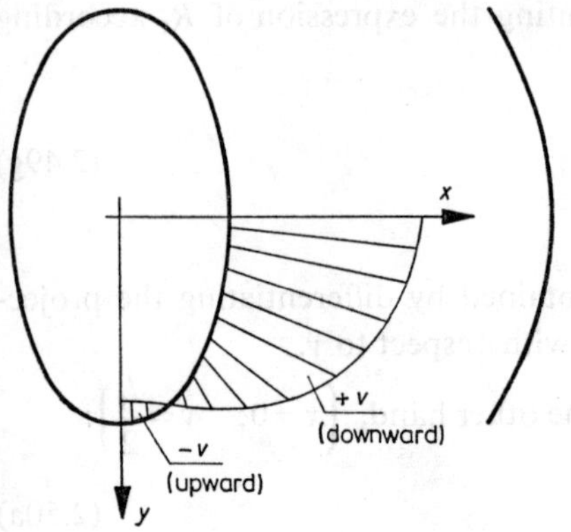

Fig. 2.10. Distribution of the support forces

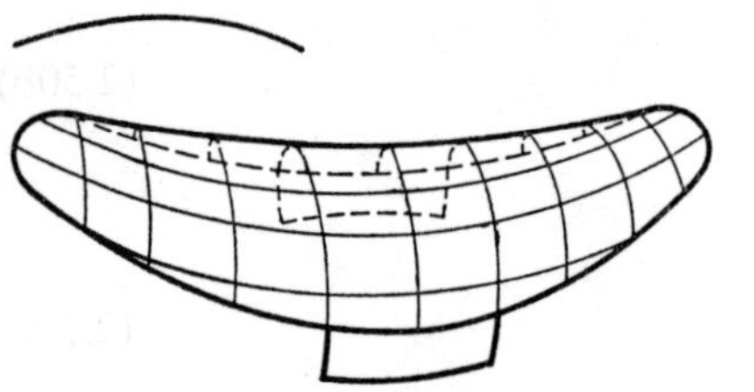

Fig. 2.11. Cable roof supported along the two lowest sections of its edge ring

the lower compression flange of which consists of the two edge rings, its upper tension flange being the net itself. In this way, a more "physical" explanation of the support force distribution described in Fig. 2.10 is obtained.

If only the two sections of the edge situated more deeply are supported, and if these supports are relatively low and rigid, then it is necessary to modify the method of calculation used so far for determining the horizontal deformation of the edge ring. Namely, in the first approximation, the edge can be regarded as being infinitely rigid in the horizontal plane because the two rigid supports prevent, to a large extent, the deformation depicted in Fig. 2.4b. One proceeds somewhat more exactly if the deformation shown in

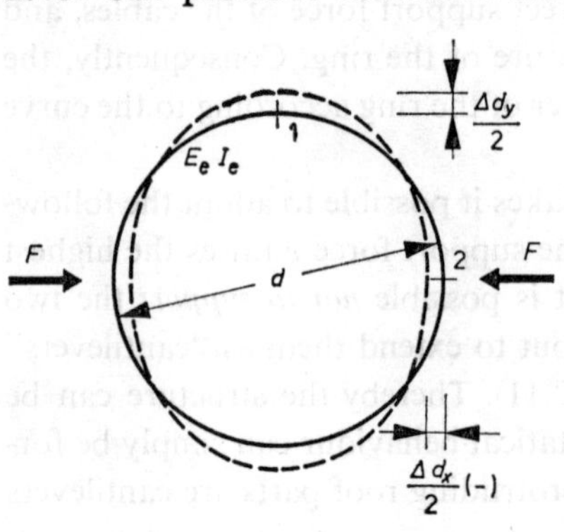

Fig. 2.12. Deformation of the edge ring due to two concentrated forces

Fig. 2.4b is corrected as indicated in Fig. 2.12, i.e. the effect of the two supports is substituted by two concentrated horizontal forces, and the horizontal displacement at the supports is set to zero. The following bending moments and displacements originate from the loading of Fig. 2.12 [13]:

$$M_1 = -0.0908\,Fd, \tag{2.51a}$$

$$M_2 = +0.1592\,Fd, \tag{2.51b}$$

$$\Delta d_x = +0.0186\,\frac{Fd^3}{E_e I_e}, \tag{2.51c}$$

$$\Delta d_y = -0.0171\,\frac{Fd^3}{E_e I_e}. \tag{2.51d}$$

The cantilever part should be dimensioned as follows:
The cable forces, the compressive force of the ring and the support forces (together with the self-weight of the ring) should be determined in the manner described above. Since, however, there is no row of columns that would carry these support forces, they should be made to act on the ring, and the deflection of the ring, as a space girder, arising from this should be determined. The ring, however, is also displaced horizontally, which in turn modifies the cable forces. Thus, once again, the ring and the cables "share" in the load-carrying.

3. Comparative Numerical Example

3.1. General Principles

To check the accuracy of the approximate method, we have calculated the internal forces of the single-layered roof in the shape of a hyperbolic paraboloid having an edge ring (Fig. 3.1) both by the approximate method described in Chapter 2 [9] and by the exact method to be discussed in Chapter 5. The approximate method — in its original, linear form — computes, for every load, the internal forces on the original net shape. Thus, the effect of the individual load types can be calculated separately and then be summed

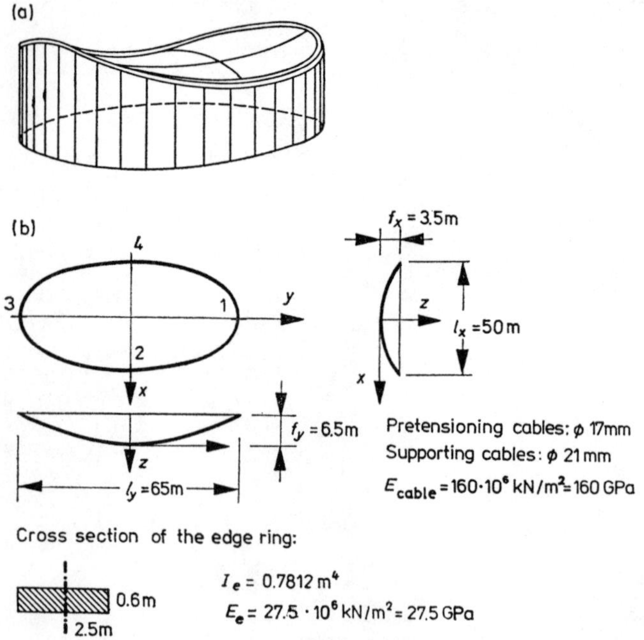

Fig. 3.1. The cable roof investigated

up as required. However, the antisymmetric loads are an exception: the deflections of the cables exhibiting inextensional (antisymmetric) deformation depend on the tensile forces acting on them due to the other previously applied loads; these loads are represented by n_x and n_y, respectively, in formulae (2.42) and (2.43). Consequently, the internal forces caused by the antisymmetric loads must be calculated separately for each combination of the other loads acting in conjunction with the antisymmetric component. In this way, it is possible to investigate the effects of the individual load types separately, as well as the magnitudes of the necessary pretensioning forces. Even so, the results involve several trials since statically indeterminate systems have to be solved; however this does not require too much computation.

The exact calculation takes the non-linear effect of the change in shape into account; hence it does not make it possible to perform separate computations for the individual load types and then to superimpose these; instead, the calculation must be made separately for every load group to be considered. The exact method can be used for checking the dimensions and internal forces adopted on the basis of the approximate calculation.

The comparative numerical example is presented as follows: the preliminary calculations, from which the dimensions are adopted, are not discussed in detail, but the determination of the internal forces due to the individual loads and the necessary pretensioning forces, computed by the approximate method, are applied to a structure having given dimensions. After this — on the basis of [9] — the results obtained by the approximate calculation on the changed net shape are briefly discussed. Finally, all these are compared with the results of the exact computation.

3.2. Basic Data

The *dimensions* of the structure are shown in Fig. 3.1. The cables are located in vertical planes at 1 m horizontal distance from each other in both directions; thus the ground plan projection of the net is a square grid. So, the data for the unit wide cable bands is as follows:

Pretensioning cable:

$$\varnothing\ 17\,\text{mm};\quad F_{\text{ult}} = 241.5\,\text{kN};\quad F_{\text{perm}} = \frac{F_{\text{ult}}}{2.8} = 86.3\,\text{kN};$$

$(EA_1)_x = 27\,700$ kN/m.

Supporting cable:

$\varnothing$ 21 mm; $F_{\text{ult}} = 368.4$ kN; $F_{\text{perm}} = 132.0$ kN; $(EA_1)_y =$
$= 42\,300$ kN/m.

The lengths of the middle pretensioning and supporting cables are given by formula (1.6) as:

$s_x = 50+0.655 = 50.655$ m,
$s_y = 65+1.735 = 66.735$ m.

The *loads* to be taken into consideration are as follows:

1. *Dead weight* (*D*): q_D=0.30 kN/m²=0.30 kPa,
2. *Pretensioning* (*P*),
3. *Total snow load* (*S*): q_S=0.80 kN/m²=0.80 kPa,
4. *Wind in the x-direction* (W_x): according to model tests, this causes a uniform suction having a shape factor $c=-0.8$. The wind pressure is p_r= =0.80 kN/m²=0.80 kPa; so $q_{\text{wind}}=-0.64$ kN/m²=−0.64 kPa.
5. *Wind in the y-direction* (W_y): by averaging on the two halves of the roof, the suction and pressure distributions obtained from wind tunnel measurements [3] made on a model of similar shape, the shape factors of Fig. 3.2 were obtained; these are divided into a uniform and an antisymmetric part. The wind acting in the 45° direction, the half-side snow load, as well as the effect of the temperature difference in the ring and the cables should all be taken into consideration, but for the sake of brevity we do not discuss these effects here; the reader is referred to Ref. [9] where results of those calculations not presented here can be found.

3.3. Internal Forces Arising from a Uniformly Distributed Vertical Load

3.3.1. Dead Load

$q = 0.30$ kN/m² = 0.30 kPa.

First we calculate the internal forces by assuming the edge ring infinitely rigid.
The load q is decomposed into components acting on the two cable rows according to formulae (2.30)–(2.31):

$q_x = 0.106$ kN/m² = 0.106 kPa,

$q_y = 0.194$ kN/m² = 0.194 kPa.

The cable forces arising from this are, according to (1.1):
$n_{xq} = -9.5$ kN/m,

$n_{yq} = +15.8$ kN/m.

Here the signs of the cable forces have been chosen to conform to the usual engineering convention. As can be seen, the uniform load causes tension in the cables in one direction and pressure in the cables in the other direction. In taking into consideration the deformation of the edge ring, the part of the cable forces causing only bending is calculated according to (2.25a):

$n_{xq}^{\mathrm{II}} = -11.8$ kN/m.

In Eq. (2.38a), which serves in calculating the deformation of the ring, firs n_{xe}^{II} is determined. For this,

according to (2.10): $d=57.1$ m,

and

according to (2.21b): $\alpha=0.876$.

Then, with these:
from (2.32): $n_{xe}^{\mathrm{II}} = -91.5\,\Delta l_x$.

According to (2.36), the internal force required for n_{xc}^{II} is $p=+1.55\,\Delta l_x$; and with this, from (2.37a), we have:

$$n_{xc} = +138\Delta l_x,$$
$$n_{yc} = +126\Delta l_x,$$

from which, according to (2.25a), the x-directional cable force causing bending only is:

$$n_{xc}^{\mathrm{II}} = +40\Delta l_x.$$

Thus, Eq. (2.38a) can be written as follows:

$$-11.8 = -91.5\Delta l_x - 40\Delta l_x,$$

and from this, the required change of diameter of the ring becomes

$$\Delta l_x = +0.0898 \text{ m}.$$

Knowing Δl_x, we are able to write down all the internal forces and deformations.
The changes in the cable forces due to Δl are given by (2.37a, b) as:

$$n_{xc} = +12.4 \text{ kN/m},$$
$$n_{yc} = +11.3 \text{ kN/m}.$$

The final cable forces are obtained as the algebraic sum of those calculated on the primary structure (with the rigid edge ring) and the forces arising from the deformation of the edge ring:

$$n_x = n_{xq}+n_{xc} = -9.5+12.4 = +2.9 \text{ kN/m},$$
$$n_y = n_{yq}+n_{yc} = +15.8+11.3 = +27.1 \text{ kN/m}.$$

Thus, by taking the deformation of the edge ring into consideration, the cable forces become tensile in both directions!
The *bending moments on the edge ring* are given by (2.12):

$$M_1 = -M_2 = +3330 \text{ kNm},$$

and the edge compressive force at the two end points of the great axis (indicating the tensile forces by a positive sign):

$$N = -n_x l_y/2 = -100 \text{ kN},$$

while at the two end points of the small axis:

$$N_2 = -n_y l_x/2 = -680 \text{ kN}.$$

The vertical displacement of the saddle point is calculated in two parts. On the primary structure with the infinitely rigid edge (subscript 0) we obtain, on the basis of (2.28) from the previously calculated value of q_x, the vertical displacement of the saddle point of the cable net:

$$w_x^0 = +0.046 \text{ m (downward)}.$$

The vertical displacement of the saddle point due to the deformation of the edge ring (subscript 1) is as follows:
The vertical displacement components arising due to the edge deformation Δl_x (and $\Delta l_y = -\alpha\Delta l_x$) are given by (2.33a) and (2.34b):

$$w_x^{\mathrm{I}} = +2.64\Delta l_x = +0.237 \text{ m},$$
$$w_x^{\mathrm{II}} = -0.67\Delta l_x = -0.060 \text{ m},$$

so that

$$w_x^1 = w_x^{\mathrm{I}} + w_x^{\mathrm{II}} = +0.177 \text{ m}.$$

Thus, the total vertical saddle point displacement due to the dead load is

$$w^{\mathrm{D}} = w_x^0 + w_x^1 = +0.046+0.177 = \underline{+0.223 \text{ m}}.$$

3.3.2. Other Uniform Loads

As it was assumed that the uniform loads could be superimposed (i.e. every change in internal force and deformation is linearly proportional to the load intensity), the internal forces and deformations caused by the other loads can be calculated from the quantities obtained from the dead load case by means of the load factors included in Table 3.1.

Table 3.1. Multiplication factors for uniformly distributed loads

Load type	Load intensity q(N/m^2=Pa)	Load factor c
Dead load	+300	+1.000
Snow	+800	+2.667
Uniform wind suction due to W_x	−640	−2.133
Uniform wind suction due to W_y	−280	−0.935

3.4. The Cable Pretensioning Required

The *minimum* pretensioning force needed is determined by the requirement that no compression should arise in the cables under any kind of load combination. Because of the influence of the deformations on the internal forces and other uncertainties, however (e.g. because of the flutter to be discussed in Section 6.3), it is not expedient just to exclude compression in the cables, but rather to prescribe "minimum" tensile forces of a certain magnitude. In the calculation of the pretensioning force it is assumed that the maximum total wind suction arising from the x-directional wind causes maximum compression in the cables. Of course, both cable rows should subsequently be checked for "cable pressure counteracting" in all loading cases.
The degree of pretensioning is "optimum" from the point of view of obtaining the maximum edge bending moment (i.e. the maximum moments are the smallest in absolute value) if the absolute values of the ring moments due to loads acting downwards and upwards (these are at the same time the maximum moments) are of the same magnitude but have opposite signs. In the case of the wind suction (load W_x) we have in the supporting cables:

$$n_y = 27.1 - 2.133 \times 27.1 = -30.7 \text{ kN/m},$$

and in the pretensioning cables:

$$n_x = +2.9 - 2.133 \times 2.9 = -3.3 \text{ kN/m}.$$

Thus, the cable force $n_{yP\min} = 30.7$ kN/m is the minimum force required in the supporting cables by the pretensioning.
The bending component of the "optimum" pretensioning force to be applied in order to obtain the most favourable bending moment distribution (see Table 3.1) is:

$$n_P^{\text{II}} = -\frac{c_{D+S} + c_{D+W_x}}{2} n_D^{\text{II}} = -\frac{3.667 - 1.133}{2} n_D^{\text{II}} = \underline{-1.267 n_D^{\text{II}}}.$$

If we apply a pretensioning force of this magnitude, edge moments having the same absolute value arise for the maximum loads acting downwards and upwards, since

$$n^{\mathrm{II}}_{D+S+P} = (3.667-1.267)\,n^{\mathrm{II}}_D = 2.4n^{\mathrm{II}}_D,$$

and

$$n^{\mathrm{II}}_{D+W_x+P} = (-1.333-1.267)\,n^{\mathrm{II}}_D = -2.4n^{\mathrm{II}}_D.$$

Thus:

$$n^{\mathrm{II}}_{xP} = -n^{\mathrm{II}}_{yP} = -1.267n^{\mathrm{II}}_{xD} = +1.267n^{\mathrm{II}}_{yD} = 1.267\times 8.2 = 10.35 \text{ kN/m}.$$

According to (2.25a), the following relationship exists between the cable force component causing bending and the total cable force:

$$n^{\mathrm{II}}_{xP} = \frac{n_{xP}\dfrac{l_y^2}{l_x^2} - n_{yP}}{1+\dfrac{l_y^2}{l_x^2}}.$$

For the forces arising in the supporting and pretensioning cables respectively, (during pretensioning) we can write, according to (2.40), that:

$$n_{yP} = n_{xP}\frac{f_x}{f_y}\frac{l_y^2}{l_x^2}.$$

Thus:

$$n^{\mathrm{II}}_{xP} = \frac{n_{xP}\left(\dfrac{l_y^2}{l_x^2} - \dfrac{f_x}{f_y}\dfrac{l_y^2}{l_x^2}\right)}{1+\dfrac{l_y^2}{l_x^2}},$$

and from this

$$n_{xP} = \frac{\left[1+\dfrac{l_y^2}{l_x^2}\right] n^{\mathrm{II}}_{xP}}{\dfrac{l_y^2}{l_x^2}\left[1-\dfrac{f_x}{f_y}\right]} = \frac{2.69}{1.69}\,\frac{10.35}{0.462} = +35.8 \text{ kN/m}.$$

From (2.40):

$$n_{yP} = +32.4 \text{ kN/m} > 30.7 \text{ kN/m} = n_{yP\,\mathrm{min}}.$$

Thus, if optimum pretensioning (from the point of view of the edge moments) is applied, a tension of 1.7 kN/m acts in the supporting cables in the case of total wind suction.

Table 3.2 contains the cable forces arising from the various combinations of the uniform loads and the pretensioning load (their resolution into components causing compression and bending is also shown):

Table 3.2. Forces arising from the various combinations of the loads

kN/m	D	P	$D+P$	$D+P+S$	$D+P+W_x$
n_x	+2.9	+35.8	+38.7	+46.4	+32.5
n_y	+27.1	+32.4	+59.5	+131.7	+1.7
n_x^{I}	+11.1	+25.45	+36.55	+66.05	+12.85
n_y^{I}	+18.9	+42.75	+61.65	+112.05	+21.35
n_x^{II}	−8.2	+10.35	+2.15	−19.65	+19.65
n_y^{II}	+8.2	−10.35	−2.15	+19.65	−19.65

Edge bending moments and edge normal forces:
On the basis of (2.12):

$$M_1 = \frac{n_{yP}^{\text{II}} d^2}{8} = -4200\ \text{kNm} = -M_2,$$

$$N_1 = -n_{xP}\frac{l_y}{2} = -1160\ \text{kN},$$

$$N_2 = -n_{yP}\frac{l_x}{2} = -810\ \text{kN}.$$

The elongations of the edge axes are directly proportional to the cable force component causing bending.
On the basis of (2.13):

$$\Delta l_x = \frac{\Delta d_x}{\sqrt{\alpha}} = \frac{1}{48}\frac{n_y^{\text{II}} d^4}{E_e I_e \sqrt{\alpha}} = 0.1098 n_y^{\text{II}} = -0.114\ \text{m},$$

and according to (2.21):

$$\Delta l_y = -0.876 \Delta l_x = +0.100\ \text{m}.$$

The vertical displacement of the saddle point
Since, according to our assumption, the x-directional pretensioning cables were prestressed, they became shorter, so that the corresponding displacement of the saddle point can be obtained from an investigation of the y-directional supporting cables. As a consequence of the edge deformation calculated in the foregoing, the centre of the middle supporting cable, in conformity with (2.33b), moves upwards by

$$w_y' = -0.181\ \text{m}.$$

The cable force $n_{yP} = +32.4$ kN/m (due to pretensioning) elongates the supporting cable. Due to this, the saddle point moves downwards, i.e. according to (2.28):

$$w_y'' = -\Delta n \frac{3}{16} \frac{l_y^2}{f_y (EA_1)_y} = +0.096 \text{ m}.$$

Thus, the final vertical saddle point displacement due to pretensioning becomes:

$$w_P = w_y' + w_y'' = -0.085 \text{ m}.$$

3.5. Investigation of the Antisymmetric Wind Load

The wind load parallel to the y-direction, i.e. to the supporting cables, (W_y), is resolved into a uniform and an antisymmetric part in Fig. 3.2. The internal forces due to the uniform load component are calculated as shown for the case of the dead load, with the multiplication factor -0.935 taken from Table 3.1.

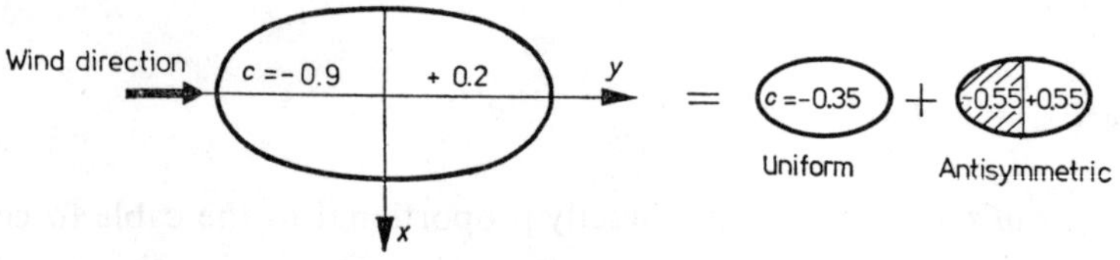

Fig. 3.2. The wind load assumed

Due to the antisymmetric load component (Fig. 3.2), the pretensioning cables change in length, and at the same time, of course, the force in them also changes. On the other hand, the supporting cables carry the load only by changing their shape; hence their cable force is not modified. (In Fig. 3.3, the direction of the cable forces is indicated; the forces acting on the edge are opposite to them.)

The deflection of the pretensioning cable at the quarter point is obtained from formula (2.44), the deflection of the supporting cable at the same point is calculated according to formula (2.43), but concerning the latter — pursuant to Section 3.1 — we must decide with which uniform loads it should be combined. Let us select the dead weight and the pretensioning as previously applied loads; then the tensile force acting in the cable as a result of the uniform loads will be as follows:

$$n_y = n_y^{D+P} + n_y^{W_y^{\text{symm}}} = +59.5 - 25.4 = 34.1 \text{ kN/m}.$$

By equating the deflections of the two cables at the quarter point, the load component acting on the pretensioning cable will be $q_x = \pm 0.397 \text{ kN/m}^2 =$

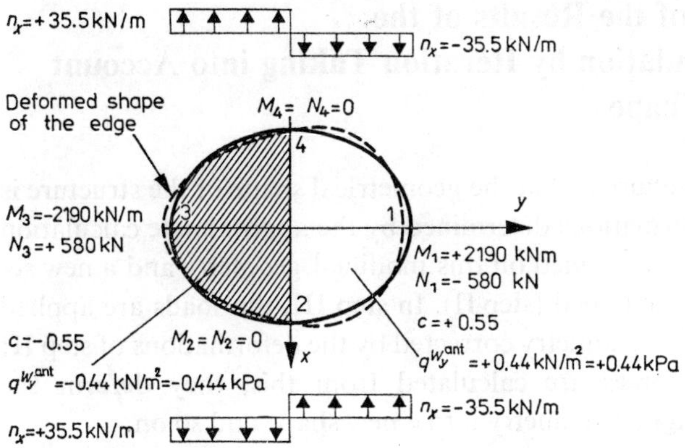

Fig. 3.3. Internal forces of the edge ring due to antisymmetric wind load

$=\pm 0.397$ kPa. The resulting excess force or force reduction, as applicable in the pretensioning cable, can be determined on the basis of the approximate relationship (2.27):

$$n_x^{W_y^{ant}} = \pm 35.5 \text{ kN/m}.$$

Edge moments and edge compressive forces due to the antisymmetric load according to (3.45a–d):

$$M_1 = -M_3 = +0.019 n_x d^2 = +2190 \text{ kNm},$$
$$M_2 = M_4 = 0,$$
$$N_1 = -N_3 = 0.288 n_x d = 580 \text{ kN},$$
$$N_2 = N_4 = 0.$$

The elongations of the axes

Since — according to our assumption — the distance between the two end points of the antisymmetrically deformed cable does not change,

$$\Delta l_x^{W_y^{ant}} = 0, \text{ and thus}$$

$$\Delta l_x^{W_y} = \Delta l_x^{W_y^{symm}} = -0.935\, \Delta l_x^D = -0.935 \times 0.090 = -0.084 \text{ m},$$

$$\Delta l_y = -\alpha \Delta l_x = +0.074 \text{ m}.$$

The vertical displacement of the saddle point:

$$w^{W_y} = w^{W_y^{symm}} + w^{W_y^{ant}} = w^{W_y^{symm}} + 0 = -0.935 \times 0.233 =$$

$$= -0.209 \text{ m (upward)}.$$

(Namely, under antisymmetric loads, the saddle point — according to our earlier approximation — is not displaced vertically.)

3.6. Improvement of the Results of the Approximate Calculation by Iteration Taking into Account the Changed Net Shape

The essence of the procedure is that the geometrical shape of the structure is corrected from the deformations determined by the approximate calculation (step I); the loads are then applied on this modified geometry and a new set of internal forces is thus obtained (step II). In step III, the loads are applied on the structure having its geometry corrected by the deformations of step II; then, again, internal forces are calculated from this, they deform the structure having the original geometry into a new shape and so on.
This iteration takes into consideration the effect of the deformation on the internal forces but, of course, it does not eliminate the error stemming from the adoption of the cable force distribution in advance.
The results corrected in this way are contained in Table 3.3. for two load combinations (dead weight + pretensioning and total snow load or total wind suction W_x, respectively).

Table 3.3. Results obtained by iteration taking modified geometry into account

Internal forces and deformations	Load group: $D+P+S$	$D+P+W_x$
n_x (kN/m)	+44.8	+33.2
n_y (kN/m)	+109.3	+6.2
M_1 (kNm)	+5280	−7650
N_1 (kN)	−1450	−1080
N_2 (kN)	−2730	−150
Δl_x (m)	+0.142	−0.205
Δl_y (m)	−0.124	+0.179
w_m (m)	+0.55	−0.308

3.7. Comparison of the Results of the Approximate Calculation with those of the Exact One

The exact calculation of the structure shown in Fig. 3.1 was carried out as described in Chapters 4 and 5. The programme was made by Miklós Berényi. This exact calculation also considered the edge ring as plane when computing the deformations. However, contrary to the approximate calculation, it did not adopt the distribution of the cable forces in advance, nor did it substitute the elliptic ring by a circle; furthermore, it always took into consideration the influence of the deformations on the internal forces, and finally, it did not consider the edge ring as infinitely rigid in the case of the antisymmetric wind load.

For the sake of a better comparison, in the case of some quantities the results of the approximate calculation corrected by iteration (i.e. by taking the changed shape into consideration), as well as the "linear result" (given by the tangent at the origin of the load-deformation curve for the exact calculation obtained by neglecting the influence of the deformation) are also indicated.

3.7.1. Pretensioning (*P*)

Applying the x- and y-directional pretensioning forces given in Section 3.4 on the cables, we determined the equilibrium shape of the net and the deformation of the edge ring (we neglected the selfweight of the net). The moments arising at the end points of the two axes of the elliptic ring and some deformation data [vertical displacement w_m of the middle (saddle) point, as well as the elongations, Δl, of the two ellipse axes] are compared with the results of the approximate calculation (Table 3.4).

3.7.2. Dead Weight (*D*) and Snow Load (*S*)

The edge ring moments and the characteristic deformation values contained in Section 3.2, due to the combination of total dead weight and snow loads (amounting to a total of 1.10 kN/m^2), are compared with the corresponding results of the approximate calculation in Table 3.4. The exact and approximate cable forces are indicated in Fig. 3.4 (the cable spacing was 5×6.50 m, i.e. 10 net-spacings were taken in both directions). In this figure the pretensioning forces were added to the cable forces caused by the dead weight and snow loads.

3.7.3. Antisymmetric (y-directional) Wind Load (W_y)

The edge moments and the characteristic deformation values detailed in Section 3.2 (Fig. 3.2) due to the y-directional wind load causing antisymmetric load and also total suction, are presented in Table 3.4 (we applied the dead weight on the structure, in addition to the wind load). The cable forces are shown in Fig. 3.5 (the effect of pretensioning is also included).

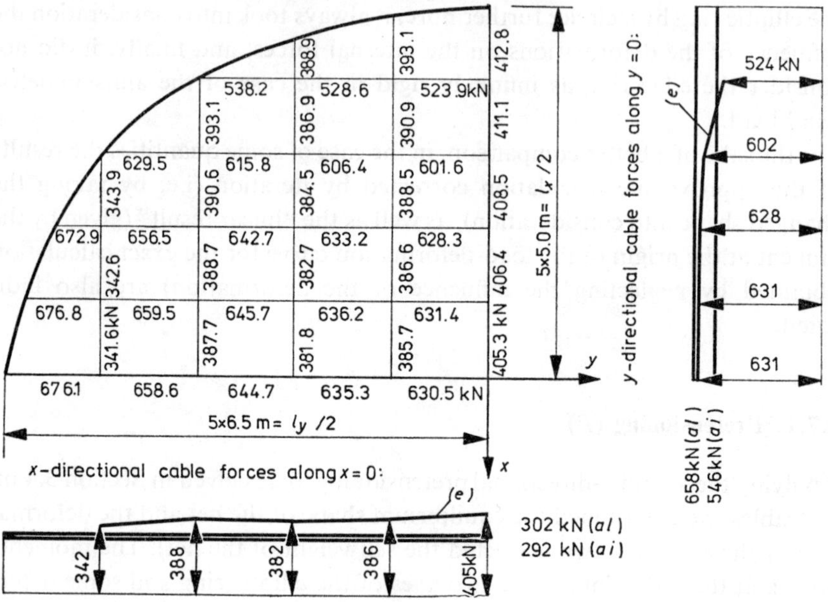

Fig. 3.4. Exact and approximate cable forces due to uniform load

3.7.4. Evaluation of the Results

The data of Table 3.4, as well as those of Figs 3.4 and 3.5, can be interpreted in the following manner:

The effect of *pretensioning* (*P*) is given by the approximate calculation with satisfactory accuracy, with the exception of the bending moments M_1 and M_2, where — due to the replacement of the elliptic ring by a circle — identical results, differing from the exact values, were obtained for both M_1 and M_2. The vertical displacement of the middle point differs for the two methods of calculation because the approximate method — in accordance with the real process of erection — considered the *x*-directional pretensioning cables as tensioned; thus they are shortened, and from this the -8.5 cm value can be calculated by relationships of geometry and strength of materials. In the approximate method however, it has not been checked whether the shape obtained in this way exactly satisfied the equilibrium condition. The exact calculation, on the other hand, leaves the mode of pretensioning out of consideration, and determines on the basis of the original net-geometry, and from the pretensioning forces (obtained from the approximate calculation) the exact equilibrium shape, which evidently differs, to a small extent, both from the initial (original) shape and from the shape obtained by the

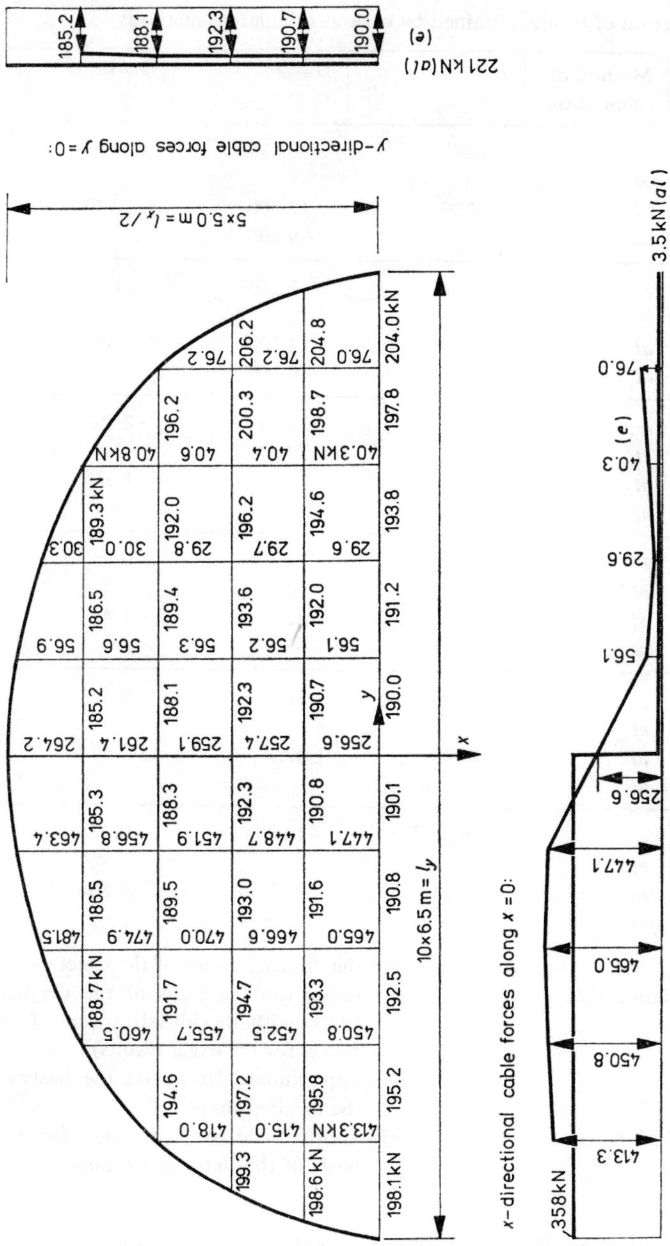

Fig. 3.5. Exact and approximate cable forces due to antisymmetric load

Table 3.4. Comparison of results obtained by various calculation methods

Internal forces and deformations	Method of calculation	P	$D+S$	$D+W_y$
M_1 (kNm)	*e*	−4682	+8 630	+2 762
	el			
	al	−4200	+12 200	+2 430
	ai		+9 500	
M_2 (kNm)	*e*	+3872	−7 270	−119
	el			
	al	+4200	−12 200	−240
	ai		−9 500	
M_3 (kNm)	*e*			−2 352
	el	$=M_1$	$=M_1$	
	al			−1 950
	ai			
w_m (cm)	*e*	+5.2	+51.8	−0.45
	el		+64.2	
	al	−8.5	+82	+1.4
	ai		+63.5	
Δl_x (cm)	*e*	−12.90	+21.9	+0.42
	el			
	al	−11.4	+32.9	+0.58
	ai			
Δl_y (cm)	*e*	+9.35	−16.8	−0.37
	el			
	al	+10.0	−28.8	−0.50
	ai			

$+w$: downward
$+M$: causing tension inside
$+\Delta l$: elongation
P = pretensioning
D = dead weight
S = snow load
W_y = y-directional wind
e = exact

el = the "linear" result of the exact calculation (on the basis of the original shape with the initial tangent of the curves for the exact results)
al = approximate, linear (on the basis of the original shape)
ai = approximate, with iteration (on the basis of the deformed shape)

approximate calculation. Thus, this is not an error inherent to the method, but is due to the initial data used.

The deviation of the results for the *uniform snow load* $(D+S)$ is caused, partly, by the substitution of the elliptic ring by a circle (M_1, M_2); the influence of the deformation on the internal forces is the other reason for the deviations. We tried to separate the latter from the other causes by calculating certain quantities by the iterative process of the approximate calculation, and also by using the initial tangent of the load-deformation curve of the exact calculation. Thus, in Table 3.4, the effect of deformation was taken into account in the same manner in values *el* and *al*; and also in *e* and *ai*. Finally, the third reason for the deviations was the arbitrary adoption of the uniform distribution of the cable forces in the approximate calculation, the error of which is shown in Fig. 3.4.

Comments similar to those for the uniform load apply to the results for the antisymmetric (y-directional) *wind load* (W_y), with the exception that, for the latter, the influence of the deformation on the internal forces is also taken into consideration in the approximate calculation, according to Section 2.4.6. Thus, the error associated with this approximation is smaller than in the case of the uniform load, so that, here, we did not make the supplementary calculations (*el* and *ai*, respectively).

Summarizing, we conclude that the approximate calculation described in Chapter 2 is suitable (even in its original, simple form, i.e. without taking into account the influence of the deformations by iteration) for an approximate estimate of the dimensions of the structure, that is, for preliminary design. It is advisable, however, to make the final computation by means of the exact method.

4. Exact Determination of the Erection Shape of the Net

4.1. Introduction

The load-bearing cable-net system is always connected to an elastic (flexible) or rigid structure. The structural members connected to the net, assumed to be elastic for the sake of simplicity, anchor and support the net, and, besides, they may also take part in the direct load-bearing and in the stiffening of the net. The cables of the net and the elastic structure interact. In a given equilibrium state of the net which can be assumed as known, the forces exerted on the structure by the net can also be assumed to be known. In accordance with this, the unloaded shape of the elastic structure — that is, its shape when it is not connected to the net — can be determined. The unstressed length of the cables of the net, or of an arbitrary element of the cables can also be determined — after releasing the connections — in the same way. Thus, when examining a possible net shape (in the cables of which arbitrarily chosen stresses are made to act so that the joint-loads of the net are unambiguously defined by the shape and by the stresses of the net), it can be assumed that it is connected to a rigid structure, the shape of which is the same as that of an elastic structure deformed by the forces transmitted from the net. If the loads, the stresses and the shape of the net change, then the system of the forces acting on the elastic structure connected to the net and, hence, its shape, also change. The change of state of the net and of the elastic structure connected to it can be analyzed only by treating these together. (They can be analyzed separately only in the initial state.) The separate analysis is suitable, at the same time, because in this way it is possible to determine the desired shape of the net by simple methods.
It is convenient to proceed with the solution of the problem in the following way. To begin with, general relationships are sought between the shape and the equilibrium of the net connected to the rigid structure. In order to keep the procedure as general as possible, not only members in tension, but also

in compression, are permitted in the net. Thus, our relationship will also be valid for bar nets, as well as cable nets.

It will be shown that the equilibrium equations of the bar net become simpler during special loadings (e.g. when only vertical joint-loads are permitted); the treatment of the individual bar forces and bar force components as free parameters enables the direct determination of the net shape corresponding to the given load. In the course of this, special attention is paid to the analysis of cable nets having a rectangular ground plan arrangement.

The simple form of the equilibrium equation for the rectangular net justifies its use in the solution of special boundary problems (non-rectangular ground plan arrangement, edge cable, interior edge cable, support boom); the formation of a net system corresponding to an orthogonal network on the net surface is then made possible by the determination of the lines of principal curvature of the surface, interpolated at the joints of the net.

The initial shape of the net is determined by the boundary, the given bar forces and joint loads. The deformation of the bar net and of the members of the structure connected to it, as well as the change of the joint loads are accompanied by changes in the joint coordinates and bar forces. The change of state of the net — on account of its significant influence on the force distribution — requires a detailed discussion, which is given in Chapter 5. In the course of this, it will be shown to what extent the differential equation describing an infinitesimally small change in the system may form the basis of an algorithm serving for the determination of a finite change of state with arbitrarily specified accuracy.

4.2. The Equilibrium Equation of a Bar Net Connected to a Rigid Structure

The bars of the net have straight axes, their cross-section is constant for a given bar, they behave elastically both in tension and compression, and their axes remain straight (in a stable state of equilibrium) during the variation of the bar force. The bars of the net are connected to each other and to the rigid structure (called "edge") by ideal, frictionless ball hinges. The bar net takes loads only at the hinges (joints).

The bar net consists of μ numbers of bars; the number of its joints is ν_0. The net is connected to the edge by ν_1 number of joints, thus the number of the real, inner net joints is $\nu=\nu_0-\nu_1$. The inner joints of the net are numbered as $j=1, 2, \ldots, \nu$, the edge joints as $j=\nu+1, \nu+2, \ldots, \nu_0$. The bars of the net are denoted by double subscripts j, k. The list of the bar symbols contains altogether μ number of pairs of j, k numbers. One variable, $\vartheta_{j,k}$, will be allocated to every possible pair of numbers j, k $(j, k = 1, 2, \dots v_0)$ using the definition given below:

$$\vartheta_{j,k} = \begin{matrix} 1 \\ 0 \end{matrix} \text{ if } j,k \begin{matrix} \text{or} \\ \text{and} \end{matrix} k,j \begin{matrix} \text{is included} \\ \text{is not included} \end{matrix} \text{ in the list of the bar symbols.}$$

The position of joint j of the net is given by the vector

$$\mathbf{r}_j = \begin{bmatrix} x_j \\ y_j \\ z_j \end{bmatrix}.$$

The length of the bar j, k is given by the positive scalar value

$$l_{j,k} = l_{k,j} = \left|\sqrt{\mathbf{r}^*_{j,k}\mathbf{r}_{j,k}}\right| = \|\mathbf{r}_{j,k}\| = \|\mathbf{r}_{k,j}\|,$$

where

$$\mathbf{r}_{j,k} = \mathbf{r}_k - \mathbf{r}_j,$$
$$\mathbf{r}_{k,j} = \mathbf{r}_j - \mathbf{r}_k = -\mathbf{r}_{j,k}.$$

The unit vector along the axis of the bar, directed from j towards k, is

$$\mathbf{e}_{j,k} = \frac{\mathbf{r}_{j,k}}{\|\mathbf{r}_{j,k}\|}.$$

The scalar value of the force associated with the bar j, k

$$s_{j,k}$$

is positive in tension, negative in compression.

In addition to the forces in the bars connected to it, the joint j of the net is subjected to the external force given by the vector $\mathbf{p}_j$. The equilibrium of the joint is expressed by the following equation (Fig. 4.1):

$$\sum_{k=1}^{0} \vartheta_{j,k} e_{j,k} s_{j,k} + \mathbf{p}_j = 0. \tag{4.1}$$

The equilibrium equation of the net joint, shown in Fig. 4.1, can also be written in the following form:

$$[\mathbf{e}_{j,k_1}\,\mathbf{e}_{j,k_2}\,\mathbf{e}_{j,k_3}\,\mathbf{e}_{j,k_4}] \begin{bmatrix} s_{j,k_1} \\ s_{j,k_2} \\ s_{j,k_3} \\ s_{j,k_4} \end{bmatrix} + \mathbf{p}_j = 0.$$

This form makes it easier to see that the equations expressing the equilibrium of every joint of the bar net can be written by a single matrix equation

$$\mathbf{G}^*\mathbf{s} + \mathbf{p} = 0. \tag{4.2}$$

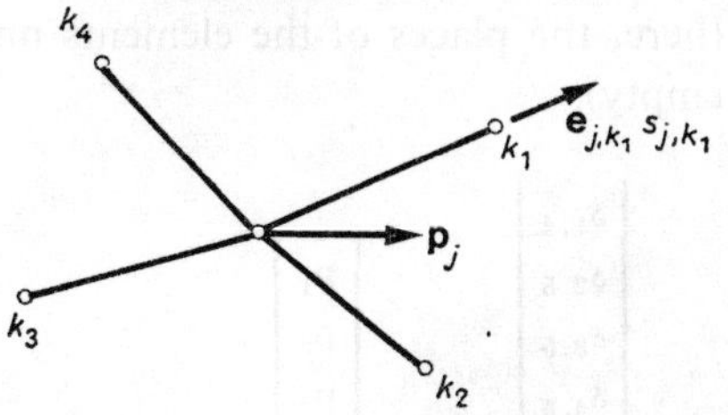

Fig. 4.1. Equilibrium of a joint of a bar net

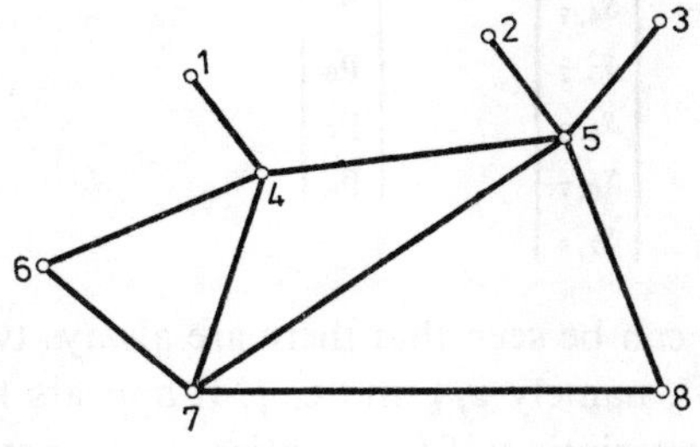

Fig. 4.2. Notations of bars and joints in the net

Here **s** is a vector of size μ, the elements of which are the scalar values $s_{j,k}$ of the bar forces; **p** is the hypervector containing ν_0 number of vectors with three elements (thus, the total number of the elements of **p** is $3\nu_0$); and $\mathbf{G}^*$ is a hypermatrix the blocks of which are column vectors with three elements (the number of its block rows is ν_0, the number of its columns μ, that is, the matrix **G** is of size $3\nu_0\mu$). The kth vector of the jth block row of matrix $\mathbf{G}^*$ is

$$\vartheta_{j,k}\mathbf{e}_{j,k}$$

that is, its vector element with index j, k is a unit vector $\mathbf{e}_{j,k}$ defining the direction of bar j, k (if the bar net has such a bar). Otherwise it is a zero vector.

Example 4.1. The content of the matrix symbols of the equation $\mathbf{G}^*\mathbf{s}+\mathbf{p}=0$, which comprises the equilibrium equations of all the joints of the bar net shown in Fig. 4.2, is:

$$\mathbf{G}^* = \begin{bmatrix} \mathbf{e}_{1,4} & & & & & & & & & \\ & \mathbf{e}_{2,5} & & & & & & & & \\ & & \mathbf{e}_{3,5} & & & & & & & \\ \mathbf{e}_{4,1} & & & \mathbf{e}_{4,5} & \mathbf{e}_{4,6} & \mathbf{e}_{4,7} & & & & \\ & \mathbf{e}_{5,2} & \mathbf{e}_{5,3} & \mathbf{e}_{5,4} & & & \mathbf{e}_{5,7} & \mathbf{e}_{5,8} & & \\ & & & & \mathbf{e}_{6,4} & & & & \mathbf{e}_{6,7} & \\ & & & & & \mathbf{e}_{7,4} & \mathbf{e}_{7,5} & & \mathbf{e}_{7,6} & \mathbf{e}_{7,8} \\ & & & & & & & \mathbf{e}_{8,5} & & \mathbf{e}_{8,7} \end{bmatrix}$$

(here, the places of the elements multiplied by the value $\vartheta_{j,k}=0$ were left empty),

$$\mathbf{s}=\begin{bmatrix} s_{1,4}\\ s_{2,5}\\ s_{3,5}\\ s_{4,5}\\ s_{4,6}\\ s_{4,7}\\ s_{5,7}\\ s_{5,8}\\ s_{6,7}\\ s_{7,8}\end{bmatrix};\quad \mathbf{p}=\begin{bmatrix}\mathbf{p}_1\\ \mathbf{p}_2\\ \mathbf{p}_3\\ \mathbf{p}_4\\ \mathbf{p}_5\\ \mathbf{p}_6\\ \mathbf{p}_7\\ \mathbf{p}_8\end{bmatrix}.$$

It can be seen that there are always two vector elements in each column of $\mathbf{G}^*$, namely $\mathbf{e}_{j,k}$ and $\mathbf{e}_{k,j}$. It appears from Eq. (4.2) of the bar net that the prescription of the position of the net joints (in other words, that of matrix $\mathbf{G}^*$) defining the geometry of the net, and of the bar force vectors, uniquely determines the forces

$$\mathbf{p}=-\mathbf{G}^*\mathbf{s}$$

to be applied on the net joints, and with which every joint is in equilibrium. In another respect the situation is no longer as unambiguous as this: with a given geometry $\mathbf{G}^*$, it is generally not possible to determine uniquely the bar forces **s** belonging to some prescribed joint load system **p** on the basis of Eq. (4.2). Namely, the bar net will be determinate both statically and kinematically only exceptionally, that is, $\mathbf{G}^*$ is only exceptionally a square matrix with non-zero determinant. Therefore, this problem has to be analyzed in a somewhat more detailed manner.

The equilibrium equation of the net joints (4.2) contains *all* the joints, including the edge points. It is evident that the forces acting on the edge joints $\mathbf{p}_j$ ($j=\nu+1,\ \ \nu+2, \ldots, \nu_0$) cannot be prescribed; these must be determined as functions of the forces acting on the inner net joints. For determining the forces acting on the edge points, the elements of the vector **p** (and simultaneously with this, the rows of the matrix $\mathbf{G}^*$) should be rearranged in such a manner that the first rows include the vectors of the forces acting on the edge joints, the other rows those acting on the inner joints. After rearrangement, Eq. (4.2) assumes the following form:

$$\begin{bmatrix}\mathbf{G}_e^*\\ \mathbf{G}_i^*\end{bmatrix}\mathbf{s}+\begin{bmatrix}\mathbf{p}_e\\ \mathbf{p}_i\end{bmatrix}=0.$$

From this equation we obtain the equilibrium equation of the inner joints of the net

$$\mathbf{G}_i^*\mathbf{s}+\mathbf{p}_i = 0 \tag{4.3}$$

on the one hand, and the relationship

$$\mathbf{p}_e = -\mathbf{G}_e^*\mathbf{s} \tag{4.4}$$

to be used in the calculation of the edge forces on the other. Thus, if the bar forces $\mathbf{s}$ due to the inner joint loads $\mathbf{p}_i$ are calculated from (4.3), it is easy to determine the edge forces $\mathbf{p}_e$ on the basis of (4.4). Accordingly, hereinafter we shall not be concerned with the problem of the edge forces separately but we shall always assume implicitly that Eq. (4.2) relates *only* to the inner joints.

The matrix of coefficients $\mathbf{G}^*$ of the equilibrium equation relating only to the inner net joints is only *exceptionally* such that $|\mathbf{G}^*|\neq 0$. In such a case, the bar net is called a statically determinate lattice structure. Only in this case is it possible to calculate the bar forces $\mathbf{s}$ directly from Eq. (4.2):

$$\mathbf{s} = -\mathbf{G}^{*-1}\mathbf{p}.$$

In the general case, $\mathbf{G}^*$ is a rectangular matrix for which we can only assume that, by the appropriate rearrangement of its columns (and simultaneously its s elements) and rows (and simultaneously its p elements), Eq. (4.2) can be generated in the following form:

$$\begin{bmatrix}\mathbf{G}_{11}^* & \mathbf{G}_{12}^*\\ \mathbf{G}_{21}^* & \mathbf{G}_{22}^*\end{bmatrix}\begin{bmatrix}\mathbf{s}_1\\ \mathbf{s}_2\end{bmatrix}+\begin{bmatrix}\mathbf{p}_1\\ \mathbf{p}_2\end{bmatrix} = 0, \tag{4.5}$$

where $\mathbf{G}_{11}^*$ is the quadratic matrix of the largest size which can be selected from $\mathbf{G}^*$, for which $|\mathbf{G}_{11}^*|\neq 0$ holds. Let us prescribe arbitrarily the bar forces included in the vector $\mathbf{s}_2$ and the joint loads contained in the vector $\mathbf{p}_1$. Then, by means of the expressions obtained from (4.5),

$$\mathbf{s}_1 = -\mathbf{G}_{11}^{*-1}\mathbf{G}_{12}^*\mathbf{s}_2-\mathbf{G}_{11}^{*-1}\mathbf{p}_1,$$

$$\mathbf{p}_2 = (\mathbf{G}_{21}^*\mathbf{G}_{11}^{*-1}\mathbf{G}_{12}^*-\mathbf{G}_{22}^*)\mathbf{s}_2+\mathbf{G}_{21}^*\mathbf{G}_{11}^{*-1}\mathbf{p}_1. \tag{4.6}$$

$\mathbf{s}_1$ and $\mathbf{p}_2$ [satisfying Eq. (4.2)] can be uniquely determined.

It follows from the foregoing that when the shape of the net (that is, the matrix $\mathbf{G}^*$) is formed arbitrarily, only a part of the joint loads can be prescribed. With the free selection of a part of the bar forces, the other part of the joint loads (i.e. those which cannot be prescribed arbitrarily) can be uniquely determined. After all, nothing prevents the calculation of a system of bar forces and joint loads which is compatible with the shape of the net. However, it is a different question to what extent the joint loads determined by the above method coincide with the loads which the net actually has to carry.

To conclude, we remark that if in (5.5) the coefficient matrix is such that

$$[\mathbf{G}_{11}^* \ \ \mathbf{G}_{12}^*]; \quad |\mathbf{G}_{11}| \neq 0,$$

then the bar net is statically indeterminate, while if its shape is

$$\begin{bmatrix} \mathbf{G}_{11}^* \\ \mathbf{G}_{21}^* \end{bmatrix},$$

the bar net is statically overdeterminate, (i.e. kinematically indeterminate).

4.3. Bar Net Loaded by a System of Parallel Forces

Let us assume that the bar net is loaded exclusively by joint forces parallel to the z-axis. In this case, the sum of the horizontal components of the bar forces gives a zero vector at every joint, independently of the magnitude of the joint load. Thus, it is possible to look for a relationship among the horizontal bar force components by making use of the projection of the net on the x, y plane. The model used for this purpose may contain only those bars which have a projection of length other than zero in the x, y plane. Such bars will be regarded as the *real* bars of the net, while any vertical bars that might exist will be treated as *fictitious* bars. The fictitious bars can be included in the model in two different ways.

(a) a distance corresponding to the length $l_{j,k}$ of the fictitious bar j, k should be prescribed between joints j and k

$$z_k = z_j + w_{j,k}; \quad w_{j,k} = l_{j,k},$$

or

(b) an excess load corresponding to the bar force should be prescribed on joints j and k:

$$\Delta p_j = s_{j,k}; \quad \Delta p_k = -s_{j,k}.$$

Equation (4.2) can be divided into a system of three scalar equilibrium equations:

$$\sum_{k=1}^{v_0} \vartheta_{j,k} \mathbf{e}_x^* \mathbf{e}_{j,k} s_{j,k} = 0,$$

$$\sum_{k=1}^{v_0} \vartheta_{j,k} \mathbf{e}_y^* \mathbf{e}_{j,k} s_{j,k} = 0, \tag{4.7}$$

$$\sum_{k=1}^{v_0} \vartheta_{j,k} \mathbf{e}_z^* \mathbf{e}_{j,k} s_{j,k} + p_j = 0, \quad (j = 1, 2, \ldots, v),$$

where $\mathbf{e}_x$, $\mathbf{e}_y$, $\mathbf{e}_z$ are unit vectors parallel to the axes x, y, z, and p_j is the scalar index number denoting the intensity of the vertical load at the jth joint. The first two equations of (4.7) can be written in the following form:

$$\sum_{k=1}^{\nu_0} \vartheta_{j,k} \cos(j, k, x) H_{j,k} = 0,$$

$$\sum_{k=1}^{\nu_0} \vartheta_{j,k} \cos(j, k, y) H_{j,k} = 0, \quad (j = 1, 2, \ldots, \nu)$$

or in a more concise matrix equation

$$\mathbf{AH} = 0, \tag{4.8}$$

where

$$\mathbf{A} = \begin{bmatrix} \mathbf{A}_x \\ \mathbf{A}_y \end{bmatrix}.$$

The elements of $\mathbf{A}_x$ and $\mathbf{A}_y$ are given by the expressions $\vartheta_{j,k} \cos(j, k, x)$ and $\vartheta_{j,k} \cos(j, k, y)$; and the elements of the vector $\mathbf{H}$ are

$$H_{j,k} = \frac{d_{j,k}}{l_{j,k}} s_{j,k}.$$

Moreover, $d_{j,k}$ is the length of the horizontal projection of bar j, k, while j, k, x and j, k, y are the angles subtended by the horizontal projection of bar j, k with the axes x and y, respectively

$$\cos(j, k, x) = \frac{l_{j,k}}{d_{j,k}} \mathbf{e}_x^* \mathbf{e}_{j,k},$$

$$\cos(j, k, y) = \frac{l_{j,k}}{d_{j,k}} \mathbf{e}_y^* \mathbf{e}_{j,k}.$$

Taking the third equation of (4.7) into consideration, and substituting the expression $\frac{z_k - z_j}{l_{j,k}}$ for $\mathbf{e}_z^* \mathbf{e}_{j,k}$; and expression $\frac{l_{j,k}}{d_{j,k}} H_{j,k}$ for $s_{j,k}$, the condition that the sum of the vertical projections of the forces acting at the inner joints be zero can be written in the form

$$\sum_{k=1}^{\nu_0} \vartheta_{j,k} \frac{z_k - z_j}{d_{j,k}} H_{j,k} + p_j = 0 \quad (j = 1, 2, \ldots, \nu). \tag{4.9}$$

It is easy to see that by assuming the scalar values p_j $(j=1, 2, \ldots, \nu)$ of the vertical forces acting on the inner joints, the scalar values $H_{j,k}$ of the horizontal components of the bar forces, and the vertical coordinates z_j $(j=\nu+1, \nu+2, \ldots, \nu_0)$ of the edge points to be known, the vertical coordinates

z_j $(j=1, 2, ..., v)$ of the inner joints can be determined. In other words: by prescribing the elevation position of the edge points, the vertical loads of the inner joints and the horizontal component of the bar forces, the shape of the net (that is, the vertical position of the inner joints) can be calculated.

To avoid any misunderstanding, it is necessary to make two remarks at this stage. Namely:

1. The selection of the index numbers $H_{j,k}$ may not be arbitrary, since the vector H formed by them has to satisfy Eq. (4.8);
2. The lengths of the horizontal projections of the bars, i.e. the quantities $d_{j,k}$ in (4.9), are regarded as given; thus in Eq. (4.9), which establishes the shape of the bar net, it is assumed that the variation of the free parameters of the equation alters only the coordinate z_j of the joints, while the horizontal projection of the net remains unchanged (of course, $l_{j,k}$ also varies with the variation of z_j).

Nothing prevents the assertion of remark 2 because the strength characteristics of the bars do not play any role in the equilibrium relationships of the net; thus it can be assumed that their length can be varied without limitation while their horizontal projection remains constant. The condition of non-zero solution of Eq. (4.8) is that $\mathbf{A}$ of order $2v\mu$ should have at least one minor of order $2v$, that is, that by the proper rearrangement of the columns and rows of $\mathbf{A}$ and $\mathbf{H}$, respectively, it should be possible to achieve such a transcription of (4.8) in which

$$\mathbf{A} = [\mathbf{A}_1 \;\; \mathbf{A}_2]; \quad \mathbf{H} = \begin{bmatrix} \mathbf{H}_1 \\ \mathbf{H}_2 \end{bmatrix},$$

and

$$|\mathbf{A}_1| \neq 0.$$

Thus, instead of (4.8)

$$[\mathbf{A}_1 \;\; \mathbf{A}_2] \begin{bmatrix} \mathbf{H}_1 \\ \mathbf{H}_2 \end{bmatrix} = 0$$

can be written, from where the expression

$$\mathbf{H}_1 = -\mathbf{A}_1^{-1} \;\; \mathbf{A}_2\mathbf{H}_2 \tag{4.10}$$

is obtained. It appears from (4.10) that if the condition concerning matrix $\mathbf{A}$ is fulfilled, then with the arbitrary prescription of elements $H_{j,k}$ contained in vector $\mathbf{H}_2$, the elements $H_{j,k}$ included in vector $\mathbf{H}_1$ can be uniquely calculated. In other words: in the case of a bar net loaded by vertical forces, the horizontal components of $\mu - 2v$ bars should be prescribed.

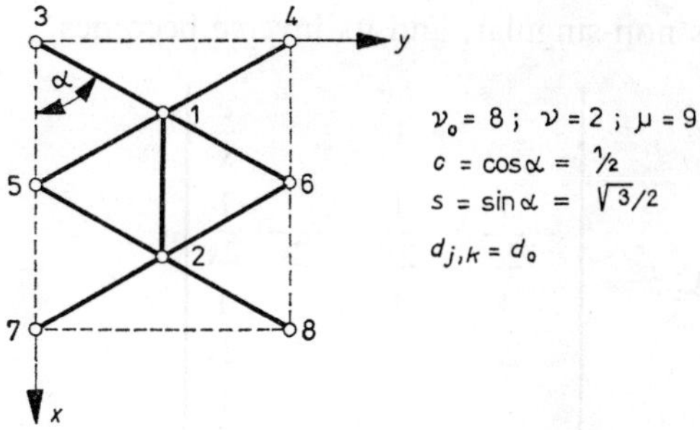

Fig. 4.3. Simple example of a bar net

Example 4.2. The inner joints (1 and 2) of the bar net, the ground plan of which is shown in Fig. 4.3, are loaded exclusively by vertical forces.
Now, Eq. (4.8), expressing that the sum of the horizontal projections equals zero, is of the following form:

$$\mathbf{AH} = \begin{bmatrix} 1 & -c & -c & c & c & & & & \\ -1 & & & & & -c & -c & c & c \\ & -s & s & -s & s & & & & \\ & & & & & -s & s & -s & s \end{bmatrix} \begin{bmatrix} H_{1,2} \\ H_{1,3} \\ H_{1,4} \\ H_{1,5} \\ H_{1,6} \\ H_{2,5} \\ H_{2,6} \\ H_{2,7} \\ H_{2,8} \end{bmatrix} = 0.$$

For the selection of the singular minor of matrix **A**, the algorithm based on dyadic decomposition, very suitable in computer calculations, is used (the details of this are discussed in the Appendix). The following solution is only an illustration.
Placing the last column of matrix **A** beside the third column, and simultaneously writing elements $H_{2,8}$ after $H_{1,4}$ in **H**, the matrix $\mathbf{A}_1$ obtained

$$\left[\begin{array}{cccc:ccccc} 1 & -c & -c & & c & c & & & \\ -1 & & & c & & & -c & -c & c \\ & -s & s & & -s & s & & & \\ & & & s & & & -s & s & -s \end{array}\right] \begin{bmatrix} H_{1,2} \\ H_{1,3} \\ H_{1,4} \\ H_{2,8} \\ H_{1,5} \\ H_{1,6} \\ H_{2,5} \\ H_{2,6} \\ H_{2,7} \end{bmatrix} = 0 = [\mathbf{A}_1 \ \ \mathbf{A}_2] \begin{bmatrix} \mathbf{H}_1 \\ \mathbf{H}_2 \end{bmatrix}$$

is non-singular, and its inverse becomes

$$\mathbf{A}_1^{-1}=\begin{bmatrix} & -1 & & \frac{c}{s}\\ -\frac{1}{2c} & -\frac{1}{2c} & -\frac{1}{2s} & \frac{1}{2s}\\ -\frac{1}{2c} & -\frac{1}{2c} & \frac{1}{2s} & \frac{1}{2s}\\ & & & \frac{1}{s}\end{bmatrix}.$$

One solution of Eq. (4.8), suitable to our example, is:

$$\begin{bmatrix}H_{1,2}\\H_{1,3}\\H_{1,4}\\H_{2,8}\end{bmatrix}=\begin{bmatrix} & & & -2c & 2c\\ & 1 & & -1 & 1\\ 1 & & & -1 & 1\\ & & 1 & -1 & 1\end{bmatrix}\begin{bmatrix}H_{1,5}\\H_{1,6}\\H_{2,5}\\H_{2,6}\\H_{2,7}\end{bmatrix}.$$

After this, let us examine how coordinates z_1 and z_2 of joints 1 and 2 of the net vary depending on the selection of the forces H.

In our example — regarding edge point coordinates z_j ($j=3, 4, ..., 8$) as known — the equation system (4.9), expressing that the sum of the vertical projections of the forces acting on the inner joints equals zero, is as follows:

$$\frac{1}{d_0}(H_{1,2}+H_{1,3}+H_{1,4}+H_{1,5}+H_{1,6})\,z_1-\frac{1}{d_0}H_{1,2}z_2=$$

$$=p_1+\frac{1}{d_0}(H_{1,3}z_3+H_{1,4}z_4+H_{1,5}z_5+H_{1,6}z_6),$$

$$\frac{1}{d_0}(H_{1,2}+H_{2,5}+H_{2,6}+H_{2,7}+H_{2,8})\,z_2-\frac{1}{d_0}H_{1,2}z_1=$$

$$=p_2+\frac{1}{d_0}(H_{2,5}z_5+H_{2,6}z_6+H_{2,7}z_7+H_{2,8}z_8).$$

(a) If, for example,

$$\mathbf{H}_2=\begin{bmatrix}H_{1,5}\\H_{1,6}\\H_{2,5}\\H_{2,6}\\H_{2,7}\end{bmatrix}=\begin{bmatrix}1\\2\\3\\4\\5\end{bmatrix},\quad\text{then}\quad \mathbf{H}_1=\begin{bmatrix}H_{1,2}\\H_{1,3}\\H_{1,4}\\H_{2,8}\end{bmatrix}=\begin{bmatrix}1\\3\\2\\4\end{bmatrix},$$

and in accordance with this, the equation system (4.9) is:

$$9z_1 - z_2 = d_0 p_1 + 3z_3 + 2z_4 + z_5 + 2z_6,$$

$$-z_1 + 17z_2 = d_0 p_2 + 3z_5 + 4z_6 + 5z_7 + 4z_8.$$

Let

$$\begin{bmatrix} p_1 \\ p_2 \end{bmatrix} = 0 \quad \text{and} \quad \begin{bmatrix} z_3 \\ z_4 \\ z_5 \\ z_6 \\ z_7 \\ z_8 \end{bmatrix} = \gamma d_0 \begin{bmatrix} 0 \\ 2 \\ 1 \\ 1 \\ 0 \\ 2 \end{bmatrix},$$

then

$$\begin{bmatrix} z_1 \\ z_2 \end{bmatrix} = \frac{\gamma d_0}{152} \begin{bmatrix} 17 & 1 \\ 1 & 9 \end{bmatrix} \begin{bmatrix} 7 \\ 15 \end{bmatrix} = \frac{\gamma d_0}{76} \begin{bmatrix} 67 \\ 71 \end{bmatrix}.$$

(b) If H_2 is chosen in a different way, e.g.

$$\mathbf{H}_2 = \begin{bmatrix} H_{1,5} \\ H_{1,6} \\ H_{2,5} \\ H_{2,6} \\ H_{2,7} \end{bmatrix} = \begin{bmatrix} 2 \\ 1 \\ 3 \\ 5 \\ 4 \end{bmatrix} \quad \text{then} \quad \mathbf{H}_1 = \begin{bmatrix} H_{1,2} \\ H_{1,3} \\ H_{1,4} \\ H_{2,8} \end{bmatrix} = \begin{bmatrix} -1 \\ 0 \\ 1 \\ 2 \end{bmatrix}$$

and

$$3z_1 + z_2 = d_0 p_1 + z_4 + 2z_5 + z_6,$$

$$z_1 + 13z_2 = d_0 p_2 + 3z_5 + 5z_6 + 4z_7 + 2z_8.$$

If, now, it is also assumed that

$$\begin{bmatrix} p_1 \\ p_2 \end{bmatrix} = 0 \quad \text{and} \quad \begin{bmatrix} z_3 \\ z_4 \\ z_5 \\ z_6 \\ z_7 \\ z_8 \end{bmatrix} = \gamma d_0 \begin{bmatrix} 0 \\ 2 \\ 1 \\ 1 \\ 0 \\ 2 \end{bmatrix},$$

then

$$\begin{bmatrix} z_1 \\ z_2 \end{bmatrix} = \frac{\gamma d_0}{38} \begin{bmatrix} 13 & -1 \\ -1 & 3 \end{bmatrix} \begin{bmatrix} 4 \\ 4 \end{bmatrix} = \frac{\gamma d_0}{76} \begin{bmatrix} 96 \\ 16 \end{bmatrix}.$$

Consequently, we see from cases (a) and (b) that, even with identical edge heights and zero vertical loads, a significantly different net shape corresponds

to the forces H selected in a different way, although the sum of the elements of H_2 was not changed.
It can easily be ensured — as will be illustrated in the next example — that **H** should have only positive elements, so that every bar is designed as a cable in tension.

Example 4.3. The joints of the lower and upper bars of the double-layer net, the ground plan and cross section of which are shown in Fig. 4.4, rest on

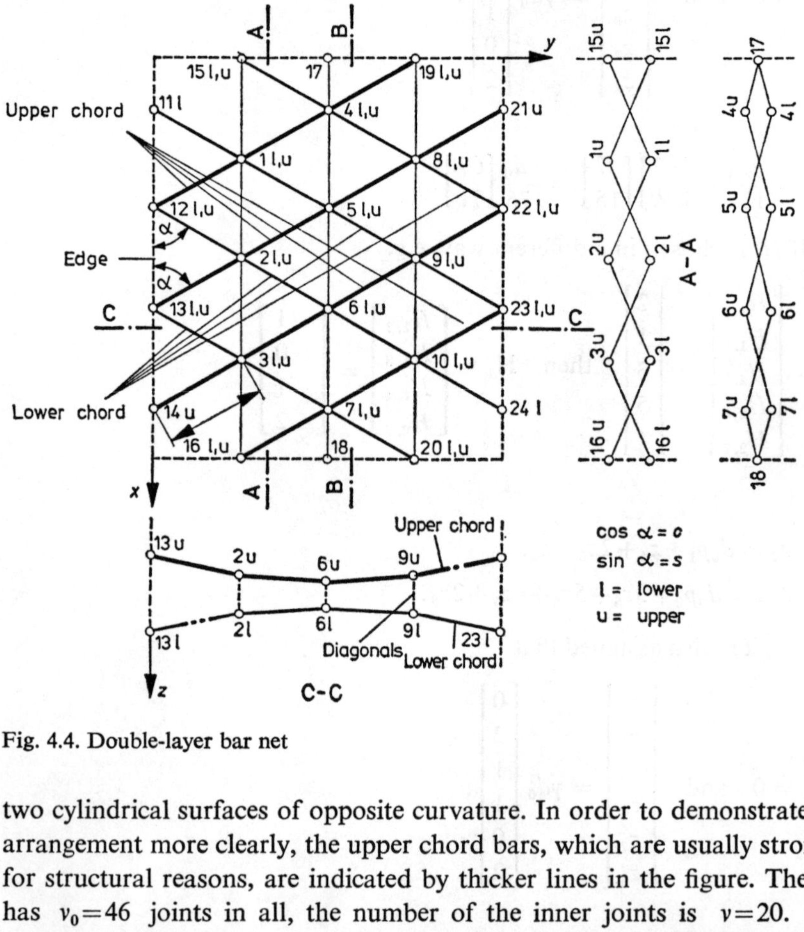

Fig. 4.4. Double-layer bar net

two cylindrical surfaces of opposite curvature. In order to demonstrate the arrangement more clearly, the upper chord bars, which are usually stronger for structural reasons, are indicated by thicker lines in the figure. The net has $\nu_0=46$ joints in all, the number of the inner joints is $\nu=20$. The number of bars is $\mu=54$. The subscripts of the bars were chosen in such a way that they should indicate the character of the bar involved: jl, $kl \doteq$ lower chord bar, ju, $ku \doteq$ upper chord bar, jl, ku, or ju, $kl \doteq$ $\doteq$ diagonal bar.
The list of bar subscripts is as follows:

1l, 2u	2l, 3u	3l, 7l	4l, 5u	5l, 6u
1l, 5l	2l, 6l	*3l, 13l*	4l, 8l	5l, 9l
1l, 11l	*2l, 12l*	3l, 16u	*4l, 15l*	5u, 6l
1l, 15u	2u, 3l	3u, 6u	*4l, 17*	5u, 8u
1u, 2l	2u, 5u	*3u, 14u*	4u, 5l	
1u, 4u	*2u, 13u*	3u, 16l	*4u, 17*	
1u, 12u			4u, 19u	
1u, 15l				
6l, 7u	7l, 18	8l, 9u	9l, 10u	10l, 20u
6l, 10l	7l, 20l	*8l, 19u*	9l, 23l	10l, 24u
6u, 7l	7u, 10u	8l, 22l	9u, 10l	10u, 20l
6u, 9u	*7u, 16u*	8u, 9l	9u, 22u	10u, 23u
	7u, 18	*8u, 19l*		
		8u, 21u		

It appears from the analysis of the coefficient matrix **A** (of order 40×54) of Eq. (4.8) related to the horizontal components of the bar forces that a non-singular minor $\mathbf{A}_1$ of order 20×20 can be selected from it. Consequently, $14(=\mu-2\nu)$ elements of **H** can be chosen freely. E.g., the elements of vector $\mathbf{H}_2$ can be chosen in such a manner that their subscripts should agree with those *set in italics* in the list of the bar subscripts.
Instead of treating mechanically the relationship between the elements of the vectors $\mathbf{H}_1$ and $\mathbf{H}_2$, let us examine more thoroughly the scalar equations of vector equation (4.8).
It is clear, from the two horizontal equilibrium equations referring to joint 1a (where according to the symbols of Fig. 4.4, $c=\cos\alpha$ and $s=\sin\alpha$):

$$H_{1l,2u}+cH_{1l,5l}-cH_{1l,11l}-H_{1l,15u}=0$$

$$sH_{1l,5l}-sH_{1l,11l}=0$$

that

$$H_{1l,5l}=H_{1l,11l},$$

and

$$H_{1l,2u}=H_{1l,15u}.$$

When moving from joint to joint, it appears that, with the adopted arrangement, the horizontal force components of chord bars lying on the same straight line in the horizontal projection are identical, independently of the joint load. The horizontal components of the bar forces of the diagonal bars, which may be considered to be elements of polygons (see, e.g., polygon 16l – 3u – 2l – 1u – 15l in Fig. 4.4, in section $A-A$), are also identical.

Thus, if 14 elements of the vector $\mathbf{H}_2$ are chosen as positive values, then all bars of the net will be in tension, and the net can also be constructed by using 14 cables with prescribed horizontal cable force components. The net outlined above will have 4 upper and 4 lower chord cables and 6 diagonal cables.

By the proper choice of the 14 H forces and the edge heights, the lower and upper "net surfaces" can be made to adopt the desired geometrical position.

4.4. Rectangular Cable Net Loaded by Vertical Forces

In the case of appropriate values of the free parameters H, the special bar net described in Example 4.3 behaves as a cable net loaded by vertical forces. By means of it, the space-enclosure having the desired shape can be created by the appropriate variation of the free parameters H and of the edge heights. The rectangular cable net to be considered next has even greater limitation from the point of view of form; yet its very simple calculation makes it useful for the solution of several different edge problems.

Hereinafter, the bar net constructed exclusively with bars in tension, and having a horizontal projection (in the x, y plane) such that its bars consist of two sets of straight lines perpendicular to each other, will be called "rectangular net". It was also assumed in the basic model of analysis that the projection of the edge points of the net on the horizontal plane lies on a closed rectangle, while the projection of the bars rests on two equi-spaced sets of straight lines, as shown in Fig. 4.5. (Note that the latter stipulation has no theoretical significance, but only practical expedience; therefore it does not alter the *essence* of the relationships to be presented.)

It follows from Eqs (4.7), limiting ourselves exclusively to vertical loads, that at every joint j, k*

$$-H_{j-1,k;\,j,k}+H_{j,k;\,j+1,k}=0,$$

$$-H_{j,k-1;\,j,k}+H_{j,k;\,j,k+1}=0,$$

* Contrary to the convention for the case of the bar net, the joints receive a double subscript in the cable net, indicating by subscripts j, k the point of intersection (in the ground plane of the jth y-directional and kth x-directional cables).

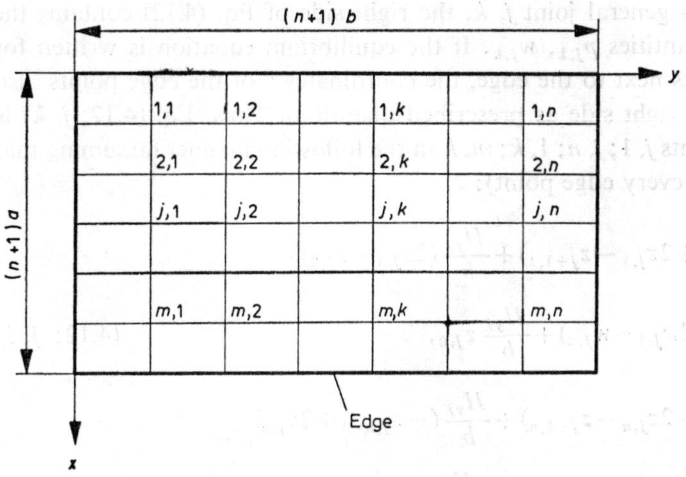

Fig. 4.5. Notations for the rectangular cable net

which means that

$$\begin{aligned} H_{j-1,k;\,j,k} &= H_{j,k;\,j+1,k} = H_{xk}; \\ H_{j,k-1;\,j,k} &= H_{j,k;\,j,k+1} = H_{yj}; \end{aligned} \quad \left\{ \begin{aligned} j &= 1,\ 2,\ \ldots,\ m \\ k &= 1,\ 2,\ \ldots,\ n \end{aligned} \right\}. \tag{4.11}$$

That is, the horizontal component of the cable force is constant along each cable.

Let us assume that at net joint j, k the x-directional cable is at height $z_{j,k}$, the y-directional cable at height $z_{j,k}+w_{j,k}$ (where $w_{j,k}$ is a positive or negative length, or possibly zero). Then, according to (4.11), (4.9) assumes now the following form:

$$\frac{z_{j-1,k}-z_{j,k}}{a}H_{xk}+\frac{z_{j+1,k}-z_{j,k}}{a}H_{xk}+\frac{(z_{j,k-1}+w_{j,k-1})-(z_{j,k}+w_{j,k})}{b}H_{yj}+$$

$$+\frac{(z_{j,k+1}+w_{j,k+1})-(z_{j,k}+w_{j,k})}{b}H_{yj}+p_{j,k}=0$$

$j = 1,\ 2,\ \ldots,\ m$

$k = 1,\ 2,\ \ldots,\ n$

$$\frac{H_{xk}}{a}(-z_{j-1,k}+2z_{j,k}-z_{j+1,k})+\frac{H_{yj}}{b}(-z_{j,k-1}+2z_{j,k}-z_{j,k+1})=$$

$$=p_{j,k}-\frac{H_{yj}}{b}(-w_{j,k-1}+2w_{j,k}-w_{j,k+1}). \tag{4.12; j, k}$$

Referring to a general joint j, k, the right side of Eq. (4.12) contains the prescribed quantities $p_{j,k}$, $w_{j,k}$. If the equilibrium equation is written for the inner joints next to the edge, the coordinates z of the edge points also appear on the right side as prescribed quantities. Thus, Eq. (4.12; j, k) is written for joints $j, 1$; j, n; $1, k$; m, k in the following manner (assuming that $w_{j,k}$ is zero at every edge point):

$$\frac{H_{x1}}{a}(-z_{j-1,1}+2z_{j,1}-z_{j+1,1})+\frac{H_{yj}}{b}(2z_{j,1}-z_{j,2}) =$$

$$= p_{j,1}-\frac{H_{yj}}{b}(2w_{j,1}-w_{j,2})+\frac{H_{yj}}{b}z_{j,0}, \qquad (4.12;\ j,\ 1)$$

$$\frac{H_{xn}}{a}(-z_{j-1,n}+2z_{j,n}-z_{j+1,n})+\frac{H_{yj}}{b}(-z_{j,n-1}+2z_{j,n}) =$$

$$= p_{j,n}-\frac{H_{yj}}{b}(-w_{j,n-1}+2w_{j,n})+\frac{H_{yj}}{b}z_{j,n+1}, \qquad (4.12;\ j,\ n)$$

$$\frac{H_{xk}}{a}(2z_{1,k}-z_{2,k})+\frac{H_{y1}}{b}(-z_{1,k-1}+2z_{1,k}-z_{1,k+1}) =$$

$$= p_{1,k}-\frac{H_{y1}}{b}(-w_{1,k-1}+2w_{1,k}-w_{1,k+1})+\frac{H_{xk}}{a}z_{0,k}, \qquad (4.12;\ 1,\ k)$$

$$\frac{H_{xk}}{a}(-z_{m-1,k}+2z_{m,k})+\frac{H_{ym}}{b}(-z_{m,k-1}+2z_{m,k}-z_{m,k+1}) =$$

$$= p_{m,k}-\frac{H_{ym}}{b}(-w_{m,k-1}+2w_{m,k}-w_{m,k+1})+\frac{H_{xk}}{a}z_{m+1,k}. \qquad (4.12;\ m,\ k)$$

In the mn number of equations (4.12; j, k), the mn number of coordinates $z_{j,k}$ ($j=1, 2, \ldots, m$; $k=1, 2, \ldots, n$) are unknown.
The system of equations from which they can be determined can also be written as a single matrix equation:

$$\frac{1}{a}\mathbf{C}_m\mathbf{Z}\mathbf{H}_x+\frac{1}{b}\mathbf{H}_y\mathbf{Z}\mathbf{C}_n = \mathbf{P}-\frac{1}{b}\mathbf{H}_y\mathbf{W}\mathbf{C}_n+\frac{1}{a}\mathbf{Z}_y\mathbf{H}_x+\frac{1}{b}\mathbf{H}_y\mathbf{Z}_x, \qquad (4.12)$$

where
$\mathbf{C}_m$ and $\mathbf{C}_n$ differ from each other only in that $\mathbf{C}_m$ is of order mm, $\mathbf{C}_n$ of order nn; otherwise

$$\mathbf{C} = \begin{bmatrix} 2 & -1 & 0 \ldots 0 \\ -1 & 2 & -1 \ldots 0 \\ 0 & -1 & 2 \ldots 0 \\ \cdots & \cdots & \cdots \\ 0 & 0 & 0 \ldots 2 \end{bmatrix}.$$

$\mathbf{H}_x$ is a diagonal matrix of order n, $\mathbf{H}_y$ is a diagonal matrix of order m

$$\mathbf{H}_x = \langle H_{x1}, H_{x2}, \dots, H_{xk}, \dots, H_{xn} \rangle,$$

$$\mathbf{H}_y = \langle H_{y1}, H_{y2}, \dots, H_{yj}, \dots, H_{ym} \rangle.$$

$\mathbf{Z}$, $\mathbf{W}$ and $\mathbf{P}$ are rectangular matrices of order mn:

$$\mathbf{Z} = \begin{bmatrix} z_{1,1} & z_{1,2} \cdots & z_{1,n} \\ z_{2,1} & z_{2,2} \cdots & z_{2,n} \\ \cdots & \cdots & \cdots \\ z_{m,1} & z_{m,2} \cdots & z_{m,n} \end{bmatrix},$$

$$\mathbf{Z}_y = \begin{bmatrix} z_{0,1} & z_{0,2} & \cdots z_{0,n} \\ 0 & 0 & \cdots 0 \\ \cdots & \cdots & \cdots \\ z_{m+1,1} & z_{m+1,2} & \cdots z_{m+1,n} \end{bmatrix},$$

$$\mathbf{Z}_x = \begin{bmatrix} z_{1,0} & 0 \dots & z_{1,n+1} \\ z_{2,0} & 0 \dots & z_{2,n+1} \\ \cdots & \cdots & \cdots \\ z_{m,0} & 0 \dots & z_{m,n+1} \end{bmatrix},$$

$$\mathbf{W} = \begin{bmatrix} w_{1,1} & w_{1,2} \cdots & w_{1,n} \\ w_{2,1} & w_{2,2} \cdots & w_{2,n} \\ \cdots & \cdots & \cdots \\ w_{m,1} & w_{m,2} \cdots & w_{m,n} \end{bmatrix}$$

$$\mathbf{P} = \begin{bmatrix} p_{1,1} & p_{1,2} \cdots & p_{1,n} \\ p_{2,1} & p_{2,2} \cdots & p_{2,n} \\ \cdots & \cdots & \cdots \\ p_{m,1} & p_{m,2} \cdots & p_{m,n} \end{bmatrix}.$$

[If the kth element of the jth row of every member in the matrix Eq. (4.12) is calculated by actually performing the assigned multiplications, the expression (4.12; j, k) is obtained.]

It is worth stressing once more that in the matrix equation (4.12) $\mathbf{C}_m$, $\mathbf{C}_n$, $\mathbf{H}_x$, $\mathbf{H}_y$, $\mathbf{P}$, $\mathbf{W}$, $\mathbf{Z}_y$, $\mathbf{Z}_x$ are *matrices containing prescribed values*, *a* and *b* are prescribed scalar values, and the unknown matrix of the equation is $\mathbf{Z}$. After some manipulation, the equation can be written in a more concise form:

$$\mathbf{A}_x\mathbf{Z}+\mathbf{Z}\mathbf{A}_y = \mathbf{G}, \tag{4.13}$$

where

$$\mathbf{A}_x = \frac{1}{a}\mathbf{H}_y^{-1}\mathbf{C}_m,$$

$$\mathbf{A}_y = \frac{1}{b}\mathbf{C}_n\mathbf{H}_x^{-1},$$

$$\mathbf{Q} = \mathbf{H}_y^{-1}\mathbf{P}\mathbf{H}_x^{-1} - \frac{1}{b}\mathbf{W}\mathbf{C}_n\mathbf{H}_x^{-1} + \frac{1}{a}\mathbf{H}_y^{-1}\mathbf{Z}_y + \frac{1}{b}\mathbf{Z}_x\mathbf{H}_x^{-1}.$$

Equation (4.13) concisely expresses the fact that, as well as prescribing the load **P** of the net, the matrix **Z** (which defines the net shape) can be influenced by the matrices $\mathbf{H}_x$, $\mathbf{H}_y$, $\mathbf{Z}_x$, $\mathbf{Z}_y$ and **W**. Of these, $\mathbf{H}_x$ and $\mathbf{H}_y$ appear on the left side of the equation as well; thus **Z** varies non-linearly as their function. If, however, $\mathbf{H}_x$ and $\mathbf{H}_y$ are taken as constant in (4.13), then the principle of superposition is valid for the other parameters, and consequently for **Z**, so that the solution of (4.13) can be produced as the sum of particular solutions. That is, the sum of the partial solutions belonging to the individual load components is equal to the solution belonging to the sum of all the loadings:

$$\mathbf{A}_x\mathbf{Z} + \mathbf{Z}\mathbf{A}_y = \mathbf{Q},$$

$$\mathbf{Q} = \mathbf{H}_y^{-1}\mathbf{P}\mathbf{H}_x^{-1} + \frac{1}{a}\mathbf{H}_y^{-1}\mathbf{Z}_y + \frac{1}{b}\mathbf{Z}_x\mathbf{H}_x^{-1} - \mathbf{W}\mathbf{A}_y,$$

$$\mathbf{Z} = \mathbf{Z}^{(1)} + \mathbf{Z}^{(2)} + \mathbf{Z}^{(3)},$$

$$\mathbf{A}_x\mathbf{Z}^{(1)} + \mathbf{Z}^{(1)}\mathbf{A}_y = \mathbf{H}_y^{-1}\mathbf{P}\mathbf{H}_x^{-1},$$

$$\mathbf{A}_x\mathbf{Z}^{(2)} + \mathbf{Z}^{(2)}\mathbf{A}_y = \frac{1}{a}\mathbf{H}_y^{-1}\mathbf{Z}_y + \frac{1}{b}\mathbf{Z}_x\mathbf{H}_x^{-1},$$

$$\mathbf{A}_x\mathbf{Z}^{(3)} + \mathbf{Z}^{(3)}\mathbf{A}_y = -\mathbf{W}\mathbf{A}_y.$$

If the advantages of writing the equilibrium equation of the rectangular cable net in the form of (4.13) are ignored, then its content can also be expressed in the form of the vector equation

$$\mathbf{A}\mathbf{z} = \mathbf{q}, \tag{4.14}$$

where **A** is an mnth order square band matrix the band width of which is $2m+1$, or $2n+1$, and **z** and **q** are vectors of dimension mn.

The forms (4.13) or (4.14) of the equilibrium equation of the rectangular cable net are equivalent, i.e. they contain the same linear equation system.

Known algorithms can be used for solving them, and these will not be discussed here. If, however, the equilibrium equation is written in the form of the matrix equation (4.13), the application of an advantageous algorithm is possible, on condition that the coefficient matrices $\mathbf{A}_x$ and $\mathbf{A}_y$ be written in the form of tern products

$$\mathbf{A}_x = \mathbf{U}_x \mathbf{\Lambda}_x \mathbf{U}_x^{-1},$$

$$\mathbf{A}_y = \mathbf{U}_y \mathbf{\Lambda}_y \mathbf{U}_y^{-1}$$

(or it may be worth producing them in such a form for the purpose of repeated use). In the tern products, $\mathbf{\Lambda}_x$ and $\mathbf{\Lambda}_y$ are diagonal matrices, whereas the other two factors are each other's inverses. By producing the coefficient matrices in the form of tern products, (4.13) is simplified in the following manner:

$$\mathbf{U}_x \mathbf{\Lambda}_x \mathbf{U}_x^{-1} \mathbf{Z} + \mathbf{Z} \mathbf{U}_y \mathbf{\Lambda}_y \mathbf{U}_y^{-1} = \mathbf{Q}.$$

If every term is multipled by $\mathbf{U}_x^{-1}$ from the left and by $\mathbf{U}_y$ from the right, we obtain

$$\mathbf{\Lambda}_x \mathbf{U}_x^{-1} \mathbf{Z} \mathbf{U}_y + \mathbf{U}_x^{-1} \mathbf{Z} \mathbf{U}_y \mathbf{\Lambda}_y = \mathbf{U}_x^{-1} \mathbf{Q} \mathbf{U}_y.$$

If the notations

$$\mathbf{D}_Q = \mathbf{U}_x^{-1} \mathbf{Q} \mathbf{U}_y \quad \text{and} \quad \mathbf{D}_z = \mathbf{U}_x^{-1} \mathbf{Z} \mathbf{U}_y,$$

are introduced, the transformed equation will assume the simple form

$$\mathbf{\Lambda}_x \mathbf{D}_z + \mathbf{D}_z \mathbf{\Lambda}_y = \mathbf{D}_Q, \tag{4.15}$$

from which the elements of $\mathbf{D}_z$ can be calculated directly from a knowledge of $\mathbf{D}_Q$. Namely, (4.15) reduces to mn number of equations with one unknown quantity each:

$$\Lambda_{x;\,j} D_{z;\,j,k} + D_{z;\,j,k} \Lambda_{y;\,k} = D_{Q;\,j,k} \quad (j = 1, 2, \ldots, m; \quad k = 1, 2, \ldots, n),$$

from which the matrix elements of the unknown $\mathbf{D}_z$ can be calculated one by one:

$$D_{z;\,j,k} = \frac{D_{Q,\,j,k}}{\Lambda_{x;\,j} + \Lambda_{y;\,k}}.$$

Having determined matrix $\mathbf{D}_z$ in this way (that is, having calculated all its elements), the required $\mathbf{Z}$ can be obtained by the transformation

$$\mathbf{Z} = \mathbf{U}_x \mathbf{D}_z \mathbf{U}_y^{-1}.$$

Several methods especially suited to computer programming are available for the production of matrices $\mathbf{A}_x$ and $\mathbf{A}_y$ in tern products, and these will not be discussed here. We mention, however, that if the horizontal components of the forces in the cables lying in the parallel planes are identical, that is, if

$$H_{xk} = H_x; \quad (k = 1, 2, ..., n),$$

$$H_{yj} = H_y; \quad (j = 1, 2, ..., n),$$

then

$$\mathbf{A}_x = \frac{1}{aH_y}\mathbf{C}_m,$$

$$\mathbf{A}_y = \frac{1}{bH_x}\mathbf{C}_n,$$

$$\mathbf{Q} = \frac{1}{H_x H_y}\mathbf{P} - \frac{1}{bH_x}\mathbf{WC}_n + \frac{1}{aH_y}\mathbf{Z}_y + \frac{1}{bH_x}\mathbf{Z}_x,$$

and in this case the elements of the matrix factors necessary for the decomposition of matrices $\mathbf{A}_y$ and $\mathbf{A}_y$ into tern products can be given directly:

$$\left.\begin{aligned} \Lambda_{x;\,j} &= \frac{4}{aH_y}\sin^2\frac{j\pi}{2(m+1)}, \\ U_{x;\,j,i} &= \sqrt{\frac{2}{m+1}}\sin\frac{ji\pi}{m+1}, \end{aligned}\right\} \quad (j, i = 1, 2, ..., m)$$

$$\left.\begin{aligned} \Lambda_{y;\,k} &= \frac{4}{bH_x}\sin^2\frac{k\pi}{2(n+1)}, \\ U_{y;\,k,h} &= \sqrt{\frac{2}{n+1}}\sin\frac{kh\pi}{n+1}. \end{aligned}\right\} \quad (k, h = 1, 2, ..., n)$$

Example 4.4. Nine x-directional and seven y-directional cables, equidistant in ground plan, are pretensioned on the rigid edge shown, in ground plan and side view, in Fig. 4.6. The horizontal components of the pretensioning forces in the cables are:

$$\mathbf{H}_x = 20\langle 1;\,1;\,1;\,1;\,1;\,1;\,1;\,1;\,1\rangle,$$

$$\mathbf{H}_y = 50\langle 0.712\ \ 0.872\ \ 0.968\ \ 1.0\ \ 0.968\ \ 0.872\ \ 0.712\rangle.$$

Vertical loads $p_{j,k}=1.2$ act on each of the inner net joints. Coordinates $z_{j,k}$ of the inner net joints are to be determined.

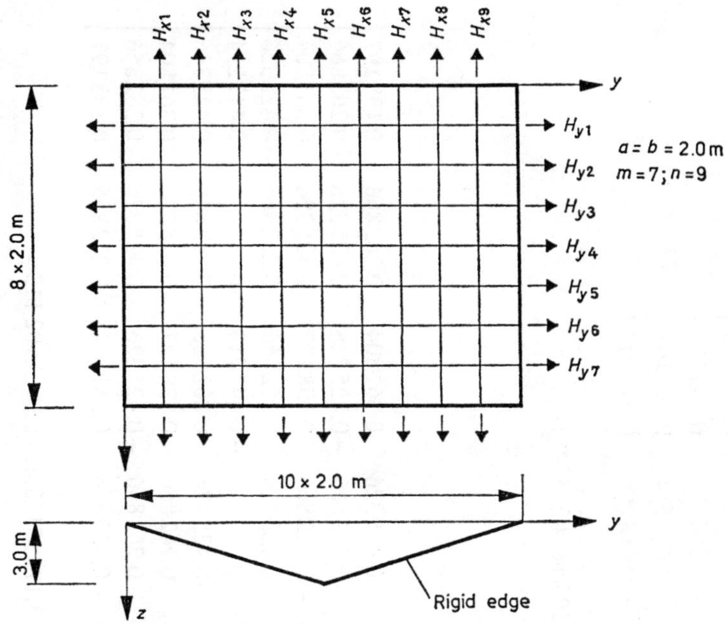

Fig. 4.6. Cable net pretensioned by constant forces

The matrices contained in the equation $\mathbf{A}_x\mathbf{Z}+\mathbf{Z}\mathbf{A}_y=\mathbf{Q}$ are:

$$\mathbf{A}_x = \frac{1}{a}\mathbf{H}_y^{-1}\mathbf{C}_m = \mathbf{U}_x\mathbf{\Lambda}_x\mathbf{U}_x^{-1},$$

$$\mathbf{A}_y = \frac{1}{b}\mathbf{C}_n\mathbf{H}_x^{-1} = \mathbf{U}_y\mathbf{\Lambda}_y\mathbf{U}_y^{-1},$$

$$\mathbf{Q} = \mathbf{H}_y^{-1}\mathbf{P}\mathbf{H}_x^{-1} - \frac{1}{b}\mathbf{W}\mathbf{C}_n\mathbf{H}_x^{-1} + \frac{1}{a}\mathbf{H}_y^{-1}\mathbf{Z}_y + \frac{1}{b}\mathbf{Z}_x\mathbf{H}_x^{-1},$$

where, for our example:

$$\mathbf{P} = 1.2\begin{bmatrix}1\\1\\1\\1\\1\\1\\1\end{bmatrix}[1;\ 1;\ 1;\ 1;\ 1;\ 1;\ 1;\ 1;\ 1]; \quad \mathbf{W} = 0; \quad \mathbf{Z}_x = 0,$$

$$\mathbf{U}_x = \begin{bmatrix} 0.141\,520 & 0.274\,052 & 0.368\,4266 & 0.415\,835 & 0.409\,768 & 0.418\,090 & 0.308\,502 \\ 0.326\,534 & 0.508\,486 & 0.432\,526 & 0.156\,662 & -0.160\,327 & -0.474\,514 & -0.397\,161 \\ 0.481\,047 & 0.421\,053 & -0.153\,008 & -0.577\,991 & -0.407\,474 & 0.266\,171 & 0.346\,315 \\ 0.541\,037 & 0 & -0.588\,346 & 0 & 0.669\,396 & 0 & -0.313\,093 \\ 0.481\,047 & -0.421\,053 & -0.153\,008 & 0.577\,991 & -0.407\,474 & -0.266\,171 & 0.346\,315 \\ 0.326\,534 & -0.508\,486 & 0.432\,526 & -0.156\,662 & -0.160\,327 & 0.474\,514 & -0.397\,161 \\ 0.141\,520 & -0.274\,052 & 0.368\,426 & -0.415\,835 & 0.409\,768 & -0.418\,090 & 0.308\,502 \end{bmatrix}$$

$$\mathbf{\Lambda}_x = \frac{1}{100} < 0.162\,972 \quad 0.681\,192 \quad 1.462\,678 \quad 2.376\,947 \quad 3.257\,686 \quad 4.110\,544 \quad 4.285\,348 >$$

$$\mathbf{U}_y = \begin{bmatrix} 0.138\,197 & 0.262\,866 & 0.361\,804 & 0.425\,326 & 0.447\,214 & 0.425\,326 & 0.361\,804 & 0.262\,866 & 0.138\,197 \\ 0.262\,866 & 0.425\,326 & 0.425\,326 & 0.262\,866 & 0 & -0.262\,866 & -0.425\,326 & -0.425\,326 & -0.262\,866 \\ 0.361\,804 & 0.425\,326 & 0.138\,197 & -0.262\,866 & -0.447\,214 & -0.262\,866 & 0.138\,197 & 0.425\,326 & 0.361\,804 \\ 0.425\,326 & 0.262\,866 & -0.262\,866 & -0.425\,326 & 0 & 0.425\,326 & 0.262\,866 & -0.262\,866 & -0.425\,326 \\ 0.447\,214 & 0 & -0.447\,214 & 0 & 0.447\,214 & 0 & -0.447\,214 & 0 & 0.447\,214 \\ 0.425\,326 & -0.262\,866 & -0.262\,866 & 0.425\,326 & 0 & -0.425\,326 & 0.262\,866 & 0.262\,866 & -0.425\,326 \\ 0.361\,804 & -0.425\,326 & 0.138\,197 & 0.262\,866 & -0.447\,214 & 0.262\,866 & 0.138\,197 & -0.425\,326 & 0.361\,804 \\ 0.262\,866 & -0.425\,326 & 0.425\,326 & -0.262\,866 & 0 & 0.262\,866 & -0.425\,326 & 0.425\,326 & -0.262\,866 \\ 0.138\,197 & -0.262\,866 & 0.361\,804 & -0.425\,326 & 0.447\,214 & -0.425\,326 & 0.361\,804 & -0.262\,866 & 0.138\,197 \end{bmatrix}$$

$$\mathbf{\Lambda}_y = \frac{1}{40} < 0.097\,887 \quad 0.381\,966 \quad 0.824\,429 \quad 1.381\,966 \quad 2.0 \quad 2.618\,034 \quad 3.175\,571 \quad 3.618\,034 \quad 3.902\,113 >

$$\mathbf{Z}_y = 0.6 \begin{bmatrix} 1 \\ 0 \\ 0 \\ 0 \\ 0 \\ 0 \\ 1 \end{bmatrix} [1;\ 2;\ 3;\ 4;\ 5;\ 4;\ 3;\ 2;\ 1].$$

Thus,

$$\mathbf{Q} = 0.12 \times 10^{-2} \begin{bmatrix} 1.404\,494 \\ 1.146\,789 \\ 1.033\,058 \\ 1.0 \\ 1.033\,058 \\ 1.146\,789 \\ 1.404\,494 \end{bmatrix} [1;\ 1;\ 1;\ 1;\ 1;\ 1;\ 1;\ 1;\ 1] +$$

$$+0.842\,696 \times 10^{-2} \begin{bmatrix} 1 \\ 0 \\ 0 \\ 0 \\ 0 \\ 0 \\ 1 \end{bmatrix} [1;\ 2;\ 3;\ 4;\ 5;\ 4;\ 3;\ 2;\ 1],$$

and

$$\mathbf{Z} = \begin{bmatrix} 0.548\,399 & 1.044\,150 & 1.479\,369 & 1.822\,927 & 1.999\,504 & \dots \\ 0.470\,458 & 0.876\,155 & 1.201\,741 & 1.423\,261 & 1.507\,432 & \dots \\ 0.413\,756 & 0.762\,674 & 1.031\,172 & 1.202\,774 & 1.262\,603 & \dots \\ 0.393\,902 & 0.723\,936 & 0.974\,926 & 1.132\,979 & 1.187\,127 & \dots \\ \dots & \dots & \dots & \dots & \dots & \dots \end{bmatrix}$$

(The dotted part contains the values corresponding to the double symmetry of the problem.)

As a check, Eq. (4.12) is written for the joint (3.4):

$$\frac{H_{x2}}{a}(-z_{2,2}+2z_{3,2}-z_{4,2})+\frac{H_{y3}}{b}(-z_{3,1}+2z_{3,2}-z_{3,3}) = p_{3,2},$$

$$10(-0.876\,155+2\times 0.762\,674-0.723\,936)+$$

$$+24.2(-0.413\,756+2\times 762\,674-1.031\,172) = 1.2.$$

4.5. Edge Condition of the Rectangular Cable Net

The matrix equation

$$\mathbf{A}_x\mathbf{Z}+\mathbf{Z}\mathbf{A}_y = \mathbf{Q}$$

determining the z coordinates of the joints of the rectangular cable net — assuming that in each cable the horizontal component (H) of the cable force is constant — is linear; thus the principle of superposition is valid for the particular solutions belonging to the components of matrix **Q** containing the loads and edge values. From this it follows that, in obtaining the spatial position of the net, by fitting it to a rigid or flexible edge, the fulfilment of the conditions on the edge is also a linear problem, in the case of constant forces H.

The determination of the net shape can be performed advantageously, from the point of view of actual calculations, by means of the algorithm described in Section 4.4, if the actual edge having an arbitrary ground plan is enclosed by a rigid edge having rectangular ground plan (i.e. by the so-called fictitious edge). The net shape, of course, has to be formed in such a manner that its elevation coordinates correspond to the requirements of space-covering inside the real edge. Care should be taken, however, that the net should approximate the desired height above the real edge as closely as possible. This can be achieved by the appropriate choice of the z coordinates of the fictitious edge and of the forces H. The net will be fitted onto the real edge only after this.

When designing the shape of the net, it is assumed that the joint loads are known. In most cases, this is the erection load, that is, the dead weight of the net applied at the joints. In principle, the cable forces and the horizontal components of these (i.e. the forces H) may be chosen arbitrarily; in practice, however, one uses available empirical data for the loads. One should not choose relatively small forces H because then the net will be loose, or it will be unserviceable because of the large sag. If high values for the forces H are chosen, they increase the cable cross-sections, which is uneconomical, and complications also arise in the anchoring. [The practical considerations governing the selection of the forces H were discussed in Section 1.2.2.3 and in Chapter 2.] When designing the shape, care should be taken also that the net remains stable not only under the dead weight, but under the service loads as well; i.e. we must ensure that sufficiently large tensile forces always act in the cables for every loading combination that is likely to arise.

The value of the joint coordinates z of the net pretensioned on a rigid edge with rectangular ground plan is governed by the following parameters:

1. the height of the edge ($\mathbf{Z}_x$, $\mathbf{Z}_y$),
2. the difference of height at the joints of the x- and y-directional cables ($\mathbf{W}$),
3. the height of those individual inner net joints ($z_{j,k}$) which are selected,
4. the horizontal components of the cable forces ($\mathbf{H}_x$, $\mathbf{H}_y$).

In order to keep the problem linear, it is desirable to fix the values of the forces H *prior to* the design of the net shape and then to use only the other three variables. The first and second of these need not be discussed separately because they appear as prescribed values in the matrix $\mathbf{Q}$ on the right side of the matrix equation of the net. The role and the consideration of the net heights prescribed at the selected points, however, should be dealt with as follows.

Frequently, it is impossible to achieve a net shape by prescribing only the edge heights and the pretensioning forces which then fix the necessary internal heights. In such cases the net can be lifted by means of one or more masts to the heights fulfilling the requirements of space enclosure. This solution, of course, can be used only if nothing prevents the location of the mast.

Let us assume that, at the net joints

$$\begin{matrix} j_1, k_1 \\ j_2, k_2 \\ \vdots \\ j_\sigma, k_\sigma \end{matrix}$$

elevation coordinates

$$z_{j_1,k_1};\quad z_{j_2,k_2};\quad \ldots;\quad z_{j_\sigma,k_\sigma}$$

are prescribed. The heights prescribed at these joints of the net can be attained by means of forces $P_1, P_2, \ldots P_\sigma$ applied at the masts. These *mast forces* can be determined from the system of equations

$$\sum_{h=1}^{\sigma} z^{(h)}_{j_i,k_i} P_h + z^{(0)}_{j_i,k_i} = z_{j_i,k_i};\quad (i = 1, 2, \ldots, \sigma),$$

where the $z^{(0)}_{j_i,k_i}$ coordinates can be determined from the equation of the basic net (i.e. without masts):

$$\mathbf{A}_x \mathbf{Z}^{(0)} + \mathbf{Z}^{(0)} \mathbf{A}_y = \mathbf{Q},$$

and the $z^{(h)}_{j_i k_i}$ coordinates can be determined from the equation of the net subject to the homogeneous edge condition, loaded by a single unit force acting only at the place of the hth mast:

$$\mathbf{A}_x \mathbf{Z}^{(h)} + \mathbf{Z}^{(h)} \mathbf{A}_y = \underset{(m)}{\mathbf{e}_{j_h}} \underset{(n)}{\mathbf{e}^*_{k_h}} \frac{1}{H_{yj_h} H_{xk_h}};\quad (h = 1, 2, \ldots, \sigma).$$

Here, $\underset{(m)}{\mathbf{e}}_{j_h}$ and $\underset{(n)}{\mathbf{e}}_{k_h}$ are the j_hth m-dimensional and k_hth n-dimensional unit vectors.

Thus, the stretching of the net on masts of prescribed heights requires as many singular solutions of the basic equation as the number of the masts, and the solution of an equation system of an order corresponding to the number of masts.

The net consisting of 13 x-directional and 15 y-directional cables stretched on the rigid edge shown in Fig. 4.7 will adopt the shape shown in Fig. 4.8 when the mast applied at point 8, 5 lifts this joint to the prescribed height $z_{8,5}$.

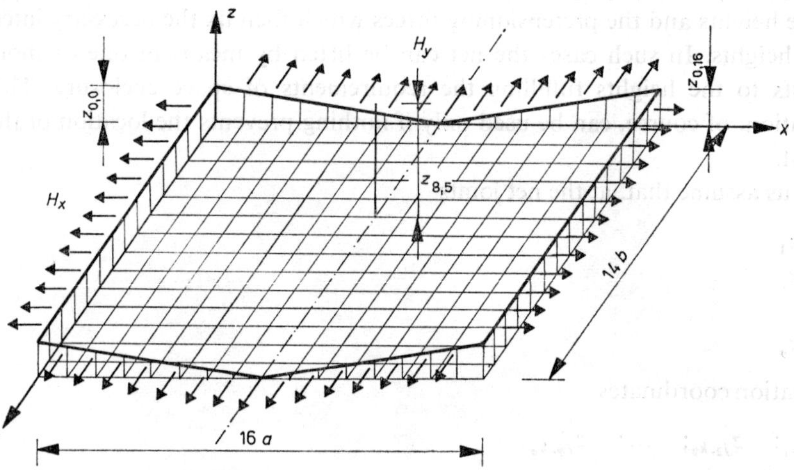

Fig. 4.7. Rectangular cable net with a mast

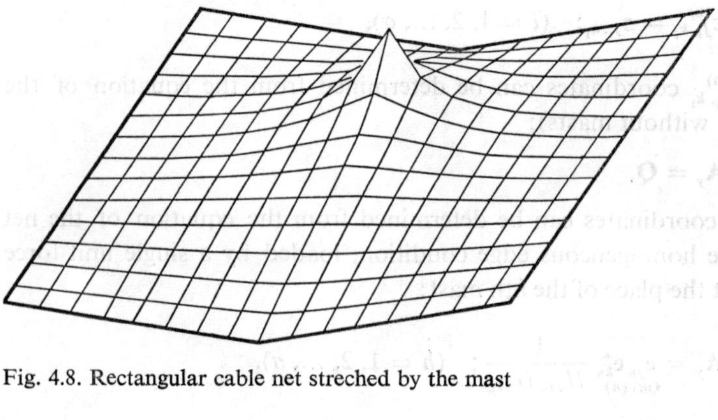

Fig. 4.8. Rectangular cable net streched by the mast

When the approximate design of the net shape has been carried out — with the aid of the rigid fictitious edge of rectangular ground plan — we have to solve the problem of exact fitting to the real edge. The investigation of two cases — possibly occurring simultaneously — will now be discussed:

1. fitting to a closed, rigid edge having an arbitrary ground plan,
2. fitting to a flexible cable edge.

4.5.1. Fitting a Rectangular Cable Net to a Closed, Rigid Edge Having an Arbitrary Ground Plan

Let us solve the *exact* fitting of the net to an edge having an arbitrary closed ground plan and height (the so-called real edge) which is *inside* an edge having a rectangular ground plan (the so-called fictitious edge) that has been taken as a basis in designing the shape. We recall that, when designing the shape of the net fitted to the fictitious edge, we could only approximately achieve the actual conditions, namely that the net should adopt the elevation position prescribed for the real edge. In general, some cables in the net will intersect the real edge higher, others lower than prescribed, since, by prescribing the elevation position of the fictitious edge, we cannot ensure that all cables will intersect the real edge exactly at the prescribed height. This can readily be seen by reference to Fig. 4.9. The net illustrated there must be fitted to the real edge at the prescribed elevation positions corresponding to points A, B, C, ..., S (thus, at 17 points in all). Although the net has 18 edge points on the fictitious edge, only the variation of 14 of these gives a linearly independent solution for the inner joint heights. Namely, matrices $\mathbf{Q}$ belonging to the variation of the height of the edge points beside the corners (e.g. edge points 1,0 and 0,1) in Eq. (4.13) differ from one another only by a constant factor. Of course, the position of the net along the real edge can also be influenced by means other than by changing the heights of the fictitious edge points. Applying a vertical force, or drawing the cables apart by a vertical distance w, can also be arbitrarily imposed on the joints in the part of the net between the real and the fictitious edge; thereby the loads within the real edge are not affected. Naturally, the fictitious loads (characterized by $p_{j,k}$ or $w_{j,k}$) to be applied at the joints of the fictitious part of the net can be taken into account only if they provide a solution linearly independent of $\mathbf{Z}$ (originating from the variation of the elevation position of the fictitious edge points). In Fig. 4.9, the vertical forces applied at points 1, 1; 1, 5; 4, 5, 4, 4: 4, 2: 4, 1; 3,1 — besides the 14 edge point displacements giving a linearly independent solution — do

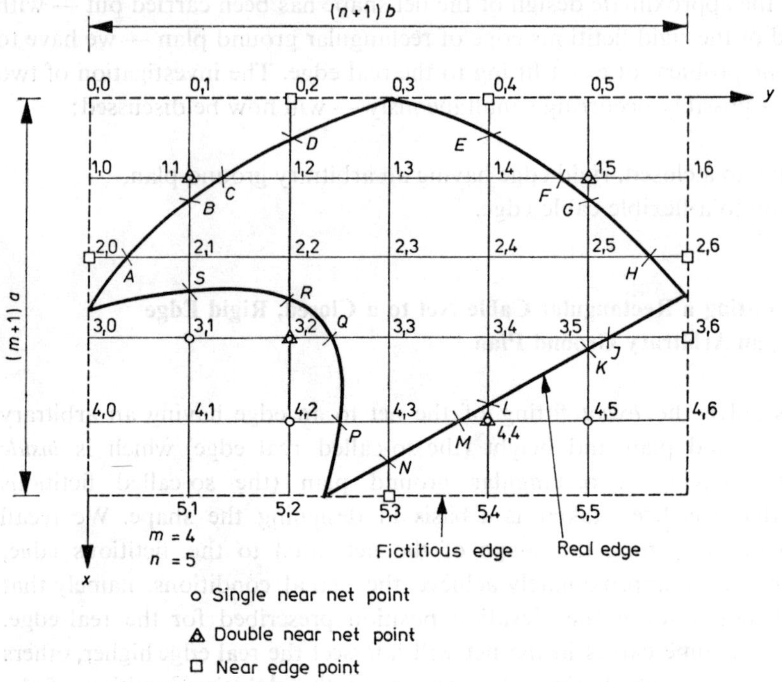

Fig. 4.9. Actual and fictitious boundary joints of a cable net

not provide a linearly independent solution; from this point of view only point 3, 2 can be taken into consideration. The drawing-apart of the cables by w, applied at points 1, 1; 1, 5; 4, 5; 4, 4; 4, 2; 4, 1 does not result into a linearly independent solution either, but w applied at points 3, 1 and 3, 2 does meet the requirement. Namely, it is easy to see that the drawing-apart of the cables at point 1,1, say, will produce a load factor

$$\mathbf{W} = \mathbf{e}_1 \mathbf{e}_1^*$$

in the expression of the matrix $\mathbf{Q}$ of Eq. (4.13); and, in accordance with this,

$$\mathbf{Q} = -\frac{1}{b}\mathbf{WCH}_x^{-1}$$

will be the linear combination of matrices $\mathbf{Q}$ corresponding to the variation of the elevation position of edge points 0, 1 and 0, 2. Thus, at 17 of the points of the real edge shown in Fig. 4.9 the edge height can be controlled by the vertical force ($p_{3,2}$) applied at point 3, 2 and by the drawing-apart of cables ($w_{3,1}$, $w_{3,2}$) at points 3,1 3, 2 in addition to the variation of height in the 14 fictitious edge points. These 17 fictitious loads (14 edge point displacements on the fictitious edge, 1 vertical force and 2 drawings-apart of cables on the fictitious part of the net) give 17 linearly independent solutions of Eq. (4.13).

It is not difficult to prove formally (but it is even simpler to see this intuitively) that the farther the point of application of the fictitious load from some point on the real edge, the smaller (and, indeed, the more rapidly decreasing) the effect it exerts on the variation of the elevation position of the point in question. Therefore, it is suitable to apply the fictitious loads on the fictitious net joints or on fictitious edge points situated nearest to the real edge points for the elevation position to be controlled. This not only provides the best solution, but the appropriate algorithm can also be constructed in the simplest manner. Let us call that joint situated on the fictitious part of the net (this can be a fictitious edge point too), which is reached on the shortest path along the cable starting from the real edge point, the "near point" of a real edge point. (In Fig. 4.9, the near point of S is point 3,1, that of R or Q is point 3, 2 and that of A is point 2, 0). Thus, we see that the same near point on the fictitious part of the net corresponds to two (possible three) points of the real edge. However, no harm ensues from this if care is taken during the process that the fictitious loads be applied on the near points in the following manner:

(a) if the near point is a fictitious edge point, then the fictitious load is a unit edge point displacement;
(b) if the near point is a joint of the fictitious net, then a unit vertical load is applied as fictitious load when this point is considered for the first time;
(c) a unit drawing-apart of the cables is applied when the same near point is considered for the second time;
(d) if the joint of the fictitious net is considered for the third time as a near point, then, instead of it, the nearest point, considered as near point only once until then, is considered once more.

The suitably selected near points of the net shown in Fig. 4.9 were allocated in accordance with the above. Summarizing the foregoing, it is advisable to tackle the problem in the following manner.

First, the points of intersection on the ground plan of the real edge and the net are determined; the number of these is v. (In the case shown in Fig. 4.9, these are points $A, B, \ldots, S$). As a second step, the joints on the fictitious net situated nearest to these points of intersection, are determined; these can either be inner or edge points. The number of single near net points is v_s, the number of double ones is ν_d, and the number of near edge points is ν_e. Of course,

$$\nu = \nu_s + 2\nu_d + \nu_e.$$

A vertical force (P) is applied on a near net point as a fictitious load, and if the point is double, then a drawing-apart of cables (W) is also applied. On a near edge point an edge-height variation (Z_x and Z_y, respectively) is applied as the fictitious load.

As a third step, the particular solution of Eq. (4.13) of the net belonging to each unit fictitious load is determined. Thus, the basic equation must be solved ν times. In each particular solution the net heights $z_\sigma^{(\tau)}$ ($\sigma, \tau = 1, 2, \ldots, \nu$) belonging to points of intersection of the net with the real edge (denoting the variation of the σth real point height due to the effect of the τth fictive load by $z_\sigma^{(\tau)}$), as well as the net heights belonging to the basic load are determined.

If the fictitious load is a vertical load applied on the net points j, k, then

$$\mathbf{P} = \mathbf{e}_j \mathbf{e}_k^*.$$

If the fictitious load is a drawing-apart of a cable, then

$$\mathbf{W} = \mathbf{e}_j \mathbf{e}_k^*$$

should be applied as the unit load in equation

$$\mathbf{A}_x \mathbf{Z} + \mathbf{Z} \mathbf{A}_y = \mathbf{Q}.$$

If the fictitious load is an edge-height variation, then one of the loads

$$\mathbf{Z}_x = \mathbf{e}_j \mathbf{e}_1^* \quad \text{(on the } y = 0 \text{ side)},$$
$$\mathbf{Z}_x = \mathbf{e}_j \mathbf{e}_n^* \quad \text{(on the } y = (n+1)b \text{ side)},$$
$$\mathbf{Z}_y = \mathbf{e}_1 \mathbf{e}_k^* \quad \text{(on the } x = 0 \text{ side)},$$
$$\mathbf{Z}_y = \mathbf{e}_m \mathbf{e}_k^* \quad \text{(on the } x = (m+1)a \text{ side)},$$

is to be applied as the unit load, depending on which side of the fictitious edge the near edge point is located.

In the fourth step, the factor X_τ of the unit fictitious loads is determined in such a manner that the total heights of the points of intersection, coming into being under the effect of the fictitious loads and the basic load, should be equal to the prescribed net height z_σ everywhere, i.e. the equation system

$$\sum_{\tau=1}^{\nu} X_\tau z_\sigma^{(\tau)} + z_\sigma^{(0)} = z_\sigma; \quad (\sigma = 1, 2, \ldots, \nu)$$

is to be solved for X_τ. If, now, the basic load of the net is completed with the sum of the unit fictitious loads multiplied by the factors X_τ, then the net will be located along the real edge, at exactly the height prescribed there.

Example 4.5. Four sides of the pentagon $A-F-G-H-J-A$ shown in ground plan in Fig. 4.10 lie on the sides of a rectangle located in the horizontal plane, while one end of the fifth side rises out of the plane by an amount z_0 (see Fig. 4.11). Let us construct the net fitting the edge having the pentagonal ground plan by means of a net consisting of three cables in each direction fitted to the fictitious rectangular edge.

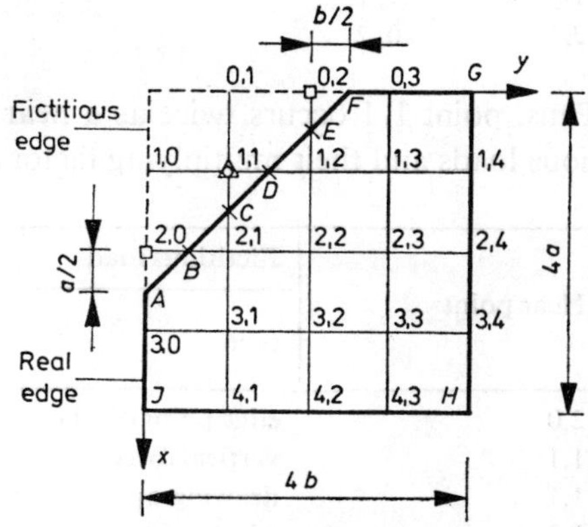

Fig. 4.10. Cable net stretched over a non-rectangular ground plan

The net has to intersect the real edge — in the section deviating from the fictitious edge — in the vertical of points B, C, D and E, at the prescribed heights. In the present case, these can be read immediately from Fig. 4.11:

$$z_B = 0.2z_0, \quad z_D = 0.6z_0,$$

$$z_C = 0.4z_0, \quad z_E = 0.8z_0.$$

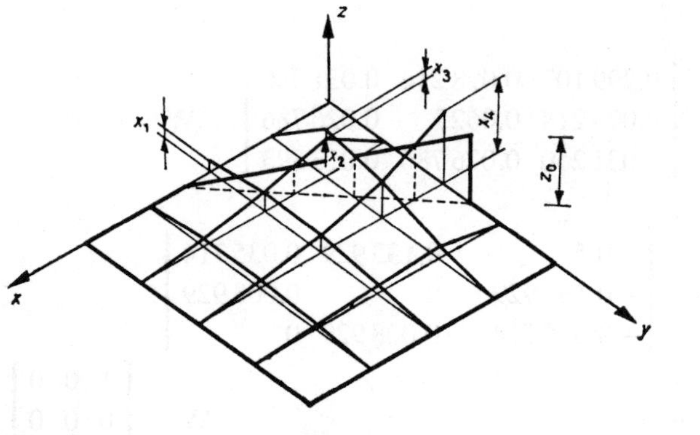

Fig. 4.11. Fictitious loads acting on the joints and on the boundary

The near points belonging to the points of intersection of the net with the real edge, are, on the fictitious part of the net (i.e. on the fictitious edge):

σ	j, k
B	2, 0
C	1, 1
D	1, 1
E	0, 2

Thus, point 1, 1 occurs twice as a near point. The corresponding unit fictitious loads and their multiplying factors are:

Near point	Fictitious load		τ
	type	multiplying factors	
2,0	edge point displacement	X_1	1
1,1	vertical force	X_2	2
1,1	drawing-apart of cables	X_3	3
0,2	edge point displacement	X_4	4

For the sake of simplicity, let us assume that in our example $\mathbf{P}=0$, $a=b=1$, $H_x=H_y=1$ and $z_0=1$.
The z coordinates of the net joints belonging to the individual fictitious loads are:

$\tau = 1$

$$\mathbf{Q} = \begin{bmatrix} 0 & 0 & 0 \\ 1 & 0 & 0 \\ 0 & 0 & 0 \end{bmatrix}; \quad \mathbf{Z}^{(1)} = \begin{bmatrix} 0.098\,214 & 0.0625 & 0.026\,786 \\ 0.330\,357 & 0.125 & 0.044\,643 \\ 0.098\,214 & 0.0625 & 0.026\,786 \end{bmatrix}; \quad \mathbf{W} = 0$$

$\tau = 2$

$$\mathbf{Q} = \begin{bmatrix} 1 & 0 & 0 \\ 0 & 0 & 0 \\ 0 & 0 & 0 \end{bmatrix}; \quad \mathbf{Z}^{(2)} = \begin{bmatrix} 0.299\,107 & 0.098\,214 & 0.031\,250 \\ 0.098\,214 & 0.0625 & 0.026\,786 \\ 0.031\,250 & 0.026\,786 & 0.013\,393 \end{bmatrix}; \quad \mathbf{W} = 0$$

$\tau = 3$

$$\mathbf{Q} = \begin{bmatrix} -2 & 1 & 0 \\ 0 & 0 & 0 \\ 0 & 0 & 0 \end{bmatrix}; \quad \mathbf{Z}^{(3)} = \begin{bmatrix} -0.5 & 0.133\,929 & 0.035\,714 \\ -0.133\,929 & 0 & 0.008\,929 \\ -0.035\,714 & -0.008\,929 & 0 \end{bmatrix};$$

$$\mathbf{W} = \begin{bmatrix} 1 & 0 & 0 \\ 0 & 0 & 0 \\ 0 & 0 & 0 \end{bmatrix}$$

$\tau = 4$

$$\mathbf{Q} = \begin{bmatrix} 0 & 1 & 0 \\ 0 & 0 & 0 \\ 0 & 0 & 0 \end{bmatrix}; \quad \mathbf{Z}^{(4)} = \begin{bmatrix} 0.098\,214 & 0.330\,357 & 0.098\,214 \\ 0.0625 & 0.125 & 0.0625 \\ 0.026\,786 & 0.044\,643 & 0.026\,786 \end{bmatrix}; \quad \mathbf{W} = 0$$

There is no displacement on the real edge due to the basic load, i.e. $z_\sigma^{(0)}=0$ ($\sigma=B, C, D, E$). With the knowledge of the net heights under the effect of the τth fictitious load — taking also into consideration the variation of the edge height prescribed at the individual fictitious loads — the height $z_\sigma^{(\tau)}$ belonging to the vertical of the σth real edge point is determined by linear interpolation. E.g.:

$z_B^{(1)}$ — height above point B (see Fig. 4.10), due to the fictitious load marked 1, of unit magnitude, which is the z-directional unit displacement of the point 2, 0 in the net

$$z_B^{(1)} = \frac{1.0+0.330\,357}{2} = 0.665\,178,$$

$$z_C^{(1)} = \frac{0.098\,214+0.330\,357}{2} = 0.214\,286.$$

$z_C^{(3)}, z_D^{(3)}$ — net heights above points C and D of the real edge, due to the unit drawing-apart of cables applied on joint 1, 1. Great attention must be paid here to the fact that the elements of matrix $\mathbf{Z}^{(3)}$ comprise the z coordinates of the joints of the x-directional cables. The data of the y-directional cables differs from these by the data of matrix $\mathbf{W}$

$$z_C^{(3)} = \frac{-0.5-0.133\,929}{2} = -0.316\,964,$$

$$z_D^{(3)} = \frac{-0.5+1.0+0.133\,929}{2} = 0.316\,964,$$

and

$$z_C^{(2)} = \frac{0.299\,107+0.098\,214}{2} = 0.198\,660,$$

$$z_D^{(2)} = \frac{0.299\,107+0.098\,214}{2} = 0.198\,660, \quad \text{etc.}$$

With the values $z_\sigma^{(\tau)}$ determined in this way we are able to write down the system of equations for the factors X_τ of the fictitious loads compatible with the height prescribed for the real edge:

$$\sum_{(\tau)} X_\tau z_\sigma^{(\tau)} + z_\sigma^{(0)} = z_\sigma; \quad \sigma = B, C, D, E$$

$$\begin{bmatrix} 0.665\,178 & 0.049\,107 & -0.066\,964 & 0.031\,250 \\ 0.214\,286 & 0.198\,660 & -0.316\,964 & 0.080\,357 \\ 0.080\,357 & 0.198\,660 & 0.316\,964 & 0.214\,286 \\ 0.031\,250 & 0.049\,107 & 0.066\,964 & 0.665\,178 \end{bmatrix} \cdot \begin{bmatrix} X_1 \\ X_2 \\ X_3 \\ X_4 \end{bmatrix} = \begin{bmatrix} 0.2 \\ 0.4 \\ 0.6 \\ 0.8 \end{bmatrix}$$

that is

$$\begin{bmatrix} X_1 \\ X_2 \\ X_3 \\ X_4 \end{bmatrix} = \begin{bmatrix} 0.143\,138\,5 \\ 1.621\,621\,9 \\ 0.120\,930\,2 \\ 1.064\,068\,8 \end{bmatrix}.$$

The z coordinates due to the joint effect of the fictitious loads formed by the factors X_τ are:

$$\mathbf{Z} = \begin{bmatrix} 0.543\,138 & 0.535\,931 & 0.163\,335 \\ 0.256\,861 & 0.252\,252 & 0.117\,411 \\ 0.088\,917 & 0.098\,806 & 0.054\,055 \end{bmatrix},$$

$$\mathbf{W} = \begin{bmatrix} 0.120\,930 & 0 & 0 \\ 0 & 0 & 0 \\ 0 & 0 & 0 \end{bmatrix}.$$

It can easily be checked that this net (see Fig. 4.11) coincides with the prescribed elevation positions at points B, C, D, E:

$$z_B = (0.143\,138 + 0.256\,861)/2 = 0.2,$$

$$z_C = (0.256\,861 + 0.543\,138)/2 = 0.4,$$

$$z_D = (0.543\,138 + 0.120\,930 + 0.535\,931)/2 = 0.6,$$

$$z_E = (0.535\,931 + 1.064\,069)/2 = 0.8.$$

The problem outlined here could have been solved with the aid of the equation system (4.12; j, k) written for points 1, 2; 1, 3; 2, 1; ...; 3, 3 as well. E.g., the equilibrium equation for point 1, 2 is

$$\frac{2H_x}{a} z_E - \frac{3H_x}{a} z_{1,2} + \frac{H_x}{a} z_{2,2} + \frac{2H_y}{b} z_D - \frac{3H_y}{b} z_{1,2} + \frac{H_y}{b} z_{1,3} = 0.$$

It is easy to check that this equation is satisfied by the above data: ($H_x/a = H_y/b = 1$)

$$2 \times 0.8 - 6 \times 0.535\,931 + 0.252\,252 + 2 \times 0.6 + 0.163\,335 = 0.000\,001.$$

4.5.2. Fitting a Rectangular Cable Net to a Flexible Edge Cable

Not only can the cables of the net be fixed to a rigid edge girder, but also to a flexible edge cable. The essential difference between these two cases is that in the design of the ground plan and the height of the rigid edge there are no restrictions, whereas the spatial position of the flexible edge cable can only be an equilibrium position corresponding to the cable forces represented by the cables fixed on it. Assuming the edge cable to be perfectly flexible, its shape will be a space polygon, at the vertices (referred to as "break points" from now on) of which the cable forces of the net and the vertical forces of prescribed value act. The equilibrium of the ith break point is expressed by the equation

$$\mathbf{e}_{i,i-1}s_{i,i-1}+\mathbf{e}_{i,i+1}s_{i,i+1}+\mathbf{k}_i = 0, \tag{4.16}$$

in which $\mathbf{k}_i$ is the force acting on the ith break point. By multiplying both sides of Eq. (4.16) by $\mathbf{e}_z^*$ and noting that

$$\mathbf{e}_z^*\mathbf{e}_{i,i-1} = \frac{z_{i-1}-z_i}{l_{i,i-1}}, \quad \text{és} \quad s_{i,i-1} = \frac{l_{i,i-1}}{d_{i,i-1}}H_{i,i-1}$$

the vertical equilibrium equation of the ith cable break point is obtained:

$$-\frac{H_{i-1,i}}{d_{i-1,i}}z_{i-1}+\left(\frac{H_{i-1,i}}{d_{i-1,i}}+\frac{H_{i,i+1}}{d_{i,i+1}}\right)z_i-\frac{H_{i,i+1}}{d_{i,i+1}}z_{i+1} = \mathbf{e}_z^*\mathbf{k}_i = P_i, \tag{4.16, z}$$

where $H_{i-1,i}$ is the horizontal component of the cable force acting in the cable section $i-1, i$, $d_{i-1,i}$ is the length of the horizontal projection of the cable section, and $P_i=\mathbf{e}_z^*\mathbf{k}_i$ is the vertical component of the force acting on the ith break point (the sum of the vertical components of the cable force and the active loading force).

It appears from Eq. (4.16, z) that if the side lengths of the horizontal projections of the cable sections (d), the horizontal components of the cable forces (H) and the vertical component of the cable force of the cables connected to the break point of the cable ($\mathbf{e}_z^*\mathbf{k}$) are known, then the z coordinate of the cable joint can be calculated.

The problem of determining the spatial position of the edge cable arises from the fact that the elevation position of the edge cable supporting the net depends on the forces of the supported cables, and that in general, in the spatial position of the cable polygon corresponding to these cable forces break points are not located on the supported net. Thus, the position of the net should be changed in such a manner that the break points of the cable polygon are located on the cables of the net so that this should be the equilibrium position of the edge cable corresponding to the supported cable forces.

This apparently complex problem can be solved in a very simple manner in the case of the rectangular net if it is assumed that the horizontal cable force components of the net are constant for each cable. A direct consequence of this assumption is that the shape of the horizontal projection of the edge cable can be determined independently of the vertical position of the net, and the problem of connecting it to the net can be solved with the knowledge of the horizontal projection of the cable. As it will be seen, this second part of the solution can be obtained without difficulty with the aid of the fictitious loads that can be applied on the joints of the fictitious part of the net.

In the next section, we shall discuss the edge cable rigidly fixed at its two ends. In special cases, the two cable end points may also coincide; then we have a closed edge cable. In the framework of Section 4.5.2.2., the inner closed edge cable will be discussed separately; its characteristic is that not even a single point on it is rigidly fixed.

4.5.2.1. Edge Cable Rigidly Fixed at its Two Ends

The horizontal component of the resultant of the forces acting on the edge cable supporting some part of the rectangular cable net pretensioned by forces having constant horizontal components in each cable can be determined in a very simple manner. The (known) horizontal components of the external forces, and hence their resultant, are independent of the vertical coordinates of the net joints. From Fig. 4.12 it is also easy to see that under a single condition (readily fulfilled) the horizontal component of the resultant of the cable forces supported by the edge cable and the projection of its line of

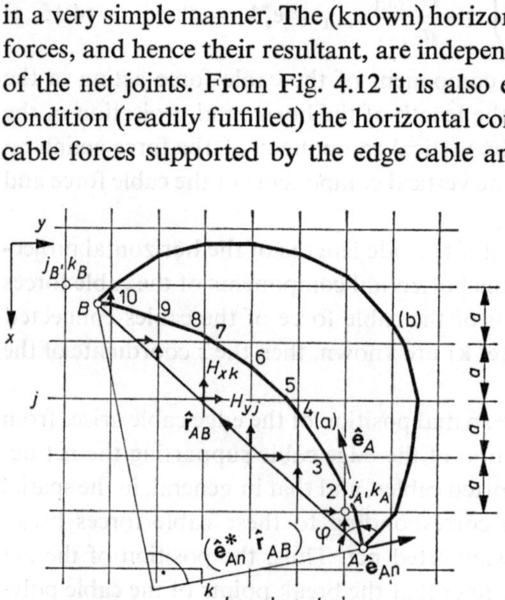

Fig. 4.12. Notations used in the investigation of the edge cable

action on the horizontal plane can be determined in advance without knowing the ground plan of the edge cable (i.e. the supported part of the net). For this it is necessary (and sufficient) to ensure that the projection of the two rigidly fixed end points of the edge cable should not fall on the projection of any cable. This requirement can easily be met even if in reality an end point of the edge cable falls on a cable in ground plan. In the latter case, the coordinates of the end point are altered by a value smaller than the accuracy prescribed for the data input and greater than the accuracy of the calculation. Thus, we achieve the condition that the end points of the edge cable do not fall on any cable in ground plan. We can then conclude that the situation shown in Fig. 4.12 is of general validity.

Let us mark the starting point of the edge cable by A, its end point by B, and let the number of the intermediate break points of the cable polygon be v. It is evident that points A and B can be connected by an infinite number of cable polygons, depending on how large an angle φ the first side of the polygon includes, with direction AB $(0<\varphi<2\pi;\ \varphi\neq\pi)$. The value of v also depends on this. [For instance, in polygon (a) of Fig. 4.12 $v=10$; in polygon (b) $v=14$.] The magnitude and line of action of the resultant, however, depend only on the cables intersected in ground plan by the straight line AB. Namely, if the polygon intersects more cables than the straight line AB does, then the former intersects the excess cables twice, and therefore such cable forces always modify the resultant by a pair of forces of zero value, i.e. no modification occurs. By prescribing the direction of the first side of the polygon, the horizontal component of the cable force belonging to each side of the cable polygon can also be uniquely determined.

Let us introduce the following notations:

j_A, k_A and j_B, k_B are the symbols of the orientation net joints of points A and B, at which

$$j_A a < x_A, \quad k_A b < y_A, \quad j_B a < x_B, \quad k_B b < y_B;$$

$\hat{\mathbf{e}}_A$ and $\hat{\mathbf{e}}_{An}$ are the unit vectors of the first side (in ground plan) of the cable polygon and the perpendicular to it, respectively (we indicate by the symbol ^ that planar vectors x, y are involved)

$\hat{\mathbf{r}}_{AB}$ is the distance vector of points A and B:

$$\hat{\mathbf{r}}_{AB} = \hat{\mathbf{r}}_B - \hat{\mathbf{r}}_A, \quad \hat{\mathbf{r}}_A = \begin{bmatrix} x_A \\ y_A \end{bmatrix}, \quad \hat{\mathbf{r}}_B = \begin{bmatrix} x_B \\ y_B \end{bmatrix}$$

v is the number of break points of the cable polygon

$\hat{\mathbf{e}}_{i-1,i}$ is the unit vector of the ith polygon side (with notation $A\equiv 0$ and $B\equiv v+1$: $\hat{\mathbf{e}}_A = \hat{\mathbf{e}}_{0,1}$)

$H_{i-1,i}$ is the horizontal component of the cable force belonging to the ith polygon side $(H_A \equiv H_{0,1})$

j_1 is the smallest of the numbers j_A and j_B.
j_2 is the greatest of the numbers j_A and j_B.
k_1 is the smallest of the numbers k_A and k_B.
k_2 is the greatest of the numbers k_A and k_B.

$\hat{\mathbf{e}}_x$ and $\hat{\mathbf{e}}_y$ are the x- and y-directional unit vectors in the plane x, y.
The cable forces intersected by the straight line AB are interpreted as pointing towards the starting direction $\hat{\mathbf{e}}_A$ (see Fig. 4.12).
The value of the horizontal cable force component $H_A \equiv H_{0,1}$ belonging to the first side of the polygon is obtained as the quotient of the sum of the moment of the intersected cable forces about point B and the perpendicular distance between $\hat{\mathbf{e}}_A$ and B:

$$H_A = H_{0,1} = \frac{\left|\sum_{j=j_1+1}^{j_2} (x_B - ja) H_{yj}\right| + \left|\sum_{k=k_1+1}^{k_2} (y_B - kb) H_{xk}\right|}{|\hat{\mathbf{e}}_A^* \hat{\mathbf{r}}_{AB}|}.$$

The vector of the horizontal component of the cable force belonging to the first polygon side is $H_A \hat{\mathbf{e}}_A$; that of the second polygon side differs from the first one by the intersected cable force. (For instance, in Fig. 4.12: $H_{1,2}\hat{\mathbf{e}}_{1,2} = H_A\hat{\mathbf{e}}_A - H_{yj_A}\hat{\mathbf{e}}_y$.) The vector of the horizontal cable force component belonging to every further polygon side can be obtained as the difference of the vector belonging to the previous side and the vector of the horizontal component of the intersected cable force. (E.g., in Fig. 4.12: $H_{2,3}\hat{\mathbf{e}}_{2,3} = H_{1,2}\hat{\mathbf{e}}_{1,2} - H_{xk_A}\hat{\mathbf{e}}_X$, $H_{3,4}\hat{\mathbf{e}}_{3,4} = H_{2,3}\hat{\mathbf{e}}_{2,3} - H_{yj_A-1}\hat{\mathbf{e}}_y$, etc.)
In Fig. 4.13 the ground plans of three cable polygons determined on the basis of the method just discussed are shown.
The starting point (A) and the end point (B) of polygons 1 and 2, shown in Fig. 4.13, are common: however, the unit vector $\hat{\mathbf{e}}_A$ of the first polygon side is different. [The direction adopted in encompassing the polygon is either clockwise ($\varrho = +1$) or counterclockwise ($\varrho = -1$).]
After this, we turn to the determination of the spatial position (height data) of the edge cable. The coordinates of the starting and end points of the edge cable $z_A \equiv z_0$, $z_B \equiv z_{\nu+1}$ (ν is the number of the break points of the cable), the height data of the fictitious edge having rectangular ground plan, and the loads acting on the joints of the real part of the net are regarded as given.
On the basis of Eq. (4.12) — or its alternative, (4.13) — the coordinates $z_{j,k}$ of the joints of the basic net, the sum of the vertical components of the cable forces acting on the individual break points, and the vertical load marked by P_i ($i = 1, 2, \dots \nu$), as well as the height coordinates $z_i^{(h)}$ of the net belonging to the vertical of the break points of the cable, are calculated.
The values $z_i^{(k)}$ of the height coordinates of the break points of the cable depend on the values P_i of the cable forces of the supported cable net. The

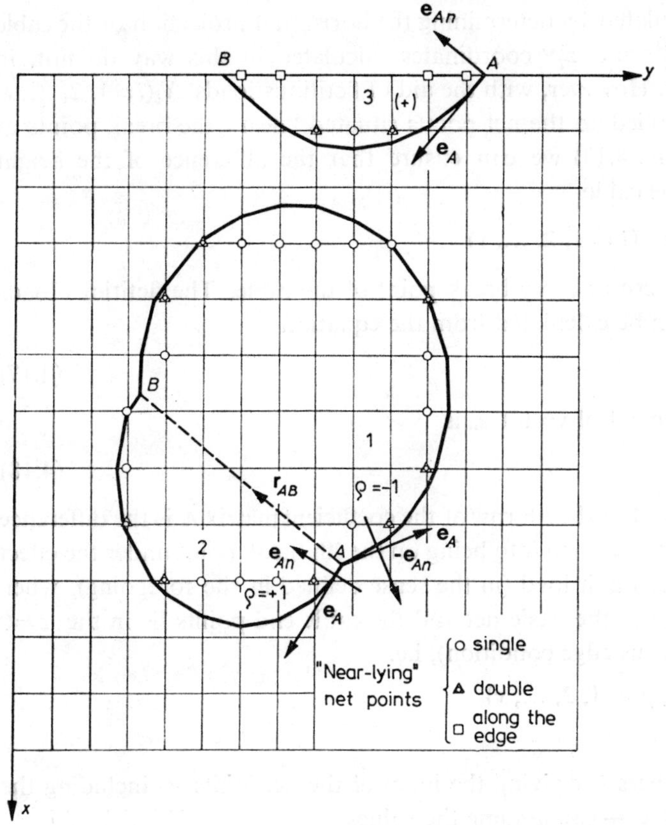

Fig. 4.13. Various types of the edge cable

equation system (4.16, z) serves for the determination of the coordinates $z_i^{(k)}$, the first equation of which is

$$\left(\frac{H_{0,1}}{d_{0,1}}+\frac{H_{1,2}}{d_{1,2}}\right) z_1^{(k)} - \frac{H_{1,2}}{d_{1,2}} z_2^{(k)} = P_1 + \frac{H_{0,1}}{d_{0,1}} z_A,$$

its ith equation:

$$-\frac{H_{i-1,i}}{d_{i-1,i}} z_{i-1}^{(k)} + \left(\frac{H_{i-1,i}}{d_{i-1,i}}+\frac{H_{i,i+1}}{d_{i,i+1}}\right) z_i^{(k)} - \frac{H_{i,i+1}}{d_{i,i+1}} z_{i+1}^{(k)} = P_i,$$

and its last (νth) equation:

$$-\frac{H_{\nu-1,\nu}}{d_{\nu-1,\nu}} z_{\nu-1}^{(k)} + \left(\frac{H_{\nu-1,\nu}}{d_{\nu-1,\nu}}+\frac{H_{\nu,\nu+1}}{d_{\nu,\nu+1}}\right) z_\nu^{(k)} = P_\nu + \frac{H_{\nu,\nu+1}}{d_{\nu,\nu+1}} z_B.$$

In the equation system, as already mentioned, z_A and z_B are given, and the length of the horizontal projection $d_{i-1,i}$ ($i=1, 2, \ldots, \nu+1$) of the polygon sides can be calculated by determining the horizontal projection of the cable polygon. The $z_i^{(h)}$ and $z_i^{(k)}$ coordinates calculated in this way do not, in general, coincide. However, with the aid of fictitious loads $X_i (i=1, 2, ..., \nu)$ which can be applied to the net points situated "near" the break points of the cable (see Fig. 4.13) we can ensure that the difference of the height coordinates of the cable

$$\Delta z_i = z_i^{(h)} - z_i^{(k)}; \quad (i = 1, 2, ..., \nu)$$

will be equal to zero at every break point of the cable. The fictitious loads marked by X_i can be calculated from the equation

$$\mathbf{A}\mathbf{x} + \mathbf{a}_0 = 0, \tag{4.17}$$

where the ith element of vector $\mathbf{a}_0$ is

$$a_{0,i} = \Delta z_i, \tag{4.18}$$

and the μth element of the ith row of the coefficient matrix $\mathbf{A}$ is the difference of the z coordinates coming into being on the ith break point under the effect of the μth fictitious unit load (in the sense defined in the foregoing), when the fictitious edge of the basic net and the cable end points lie in the $z=0$ plane (homogeneous edge condition), i.e.

$$a_{i,\mu} = \Delta z_{i,\mu}; \quad (i, \mu = 1, 2, ..., \nu)$$

$$\mathbf{A} = [a_{i,\mu}].$$

The pair of numbers j, k giving the index of the net joints — including the fictitious rigid edge — can assume the values

$$j = 0, 1, 2, ..., m, m+1$$

$$k = 0, 1, 2, ..., n, n+1$$

If $j=0$, or $j=m+1$, or $k=0$, or $k=n+1$, the points lying on the fictitious edge are involved. The fictitious load corresponding to this is a unit vertical edge point displacement:

when $j = 0$ $\quad \mathbf{P} = \frac{H_{x,k}}{a} \underset{(m)(n)}{\mathbf{e}_1 \mathbf{e}_k^*}$

when $j = m+1$ $\quad \mathbf{P} = \frac{H_{x,k}}{a} \underset{(m)(n)}{\mathbf{e}_m \mathbf{e}_k^*}$

when $k = 0$ $\quad \mathbf{P} = \frac{H_{y,j}}{b} \underset{(m)(n)}{\mathbf{e}_j \mathbf{e}_1^*}$

when $k = n+1$ $\quad \mathbf{P} = \frac{H_{y,j}}{b} \underset{(m)(n)}{\mathbf{e}_j \mathbf{e}_n^*}$

(here $\mathbf{e}_q$ is the qth unit vector of dimension m, or n).

If an edge point is not involved, then the fictitious load, when the index is used for the first time, is:

$$\mathbf{P} = \underset{(m)(n)}{\mathbf{e}_j \mathbf{e}_k^*},$$

and when used for the second time:

$$\mathbf{W} = \underset{(m)(n)}{\mathbf{e}_j \mathbf{e}_k^*}.$$

The elements of vector x obtained by the solution of Eq. (4.17) are the multiplying factors for the fictitious loads. The net and cable heights coming into being under the joint effect of the basic load and the fictitious loads formed by these multiplying factors will coincide at every break point of the cable.

There are four-edge cables in the net shown in our next example, but because of double symmetry, we can treat it as if there were only a single-edge cable.

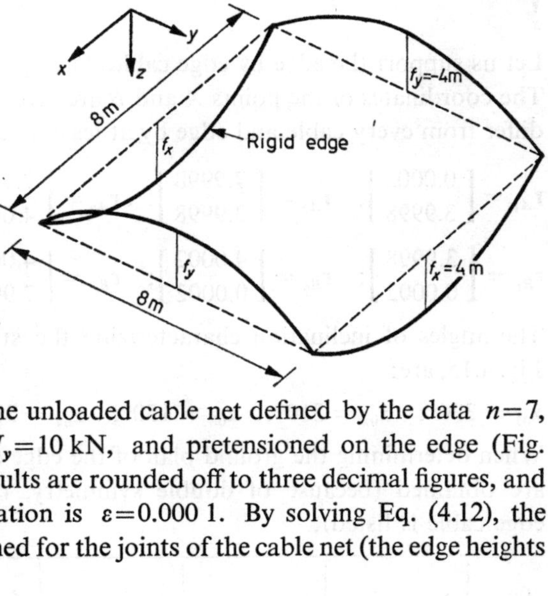

Fig. 4.14. Curved rigid edge of a rectangular cable net

Example 4.6. Let us take the unloaded cable net defined by the data $n=7$, $m=7$, $a=b=1$ m, $H_x=H_y=10$ kN, and pretensioned on the edge (Fig. 4.14) as a basic net. The results are rounded off to three decimal figures, and the accuracy of the calculation is $\varepsilon=0.000\,1$. By solving Eq. (4.12), the following heights are obtained for the joints of the cable net (the edge heights are also indicated):

0.000	−1.750	−3.000	−3.750	−4.000	−3.750	−3.000	−1.750	0.000
1.750	−0.000	−1.250	−2.000	−2.250	−2.000	−1.250	−0.000	1.750
3.000	1.250	−0.000	−0.750	−1.000	−0.750	0.000	1.250	3.000
3.750	2.000	0.750	−0.000	−0.250	0.000	0.750	2.000	3.750
4.000	2.250	1.000	0.250	−0.000	0.250	1.000	2.250	4.000
3.750	2.000	0.750	−0 000	−0.250	−0.000	0.750	2.000	3.750
3.000	1.250	−0.000	−0.750	−1.000	−0.750	−0.000	1.250	3.000
1.750	0.000	−1.250	−2.000	−2.250	−2.000	−1.250	−0 000	1.750
0.000	−1.750	−3.000	−3.750	−4.000	−3.750	−3.000	−1.750	0.000

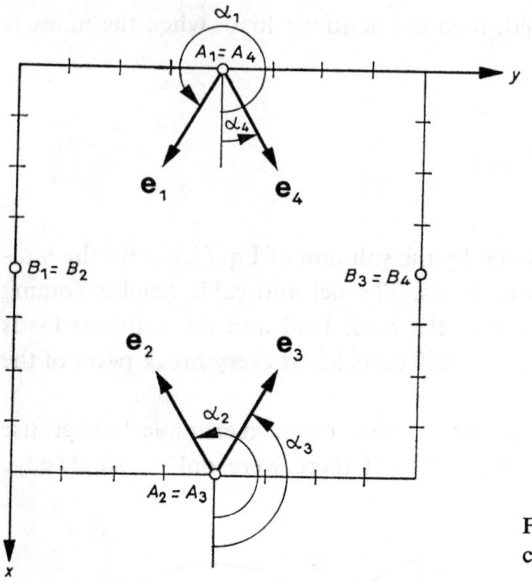

Fig. 4.15. Starting data for the construction of the edge cable

Let us support the edge by edge cables having the data shown in Fig. 4.15. The coordinates of the points A and B are given in such a manner that they differ from every cable and edge by at least a value of 2ε:

$$\mathbf{r}_{A_1} = \begin{bmatrix} 0.0002 \\ 3.9998 \end{bmatrix}; \quad \mathbf{r}_{A_2} = \begin{bmatrix} 7.9998 \\ 3.9998 \end{bmatrix}; \quad \mathbf{r}_{A_3} = \begin{bmatrix} 7.9998 \\ 4.0002 \end{bmatrix}; \quad \mathbf{r}_{A_4} = \begin{bmatrix} 0.0002 \\ 4.0002 \end{bmatrix};$$

$$\mathbf{r}_{B_1} = \begin{bmatrix} 3.9998 \\ 0.0002 \end{bmatrix}; \quad \mathbf{r}_{B_2} = \begin{bmatrix} 4.0002 \\ 0.0002 \end{bmatrix}; \quad \mathbf{r}_{B_3} = \begin{bmatrix} 4.0002 \\ 7.9998 \end{bmatrix}; \quad \mathbf{r}_{B_4} = \begin{bmatrix} 3.9998 \\ 7.9998 \end{bmatrix}.$$

The angles of inclination characterizing the starting directions, shown in Fig. 4.15, are:

$\alpha_{01} = 330°$; $\alpha_{02} = 210°$; $\alpha_{03} = 150°$; $\alpha_{04} = 30°$.

When determining the ground plan of the edge cables, the following values are obtained (because of double symmetry, only the data for the first edge cable is listed):

i	x_i	y_i	$H_{i,i+1}$	$d_{i,i+1}$	Subscript of the near point	
0	0.0002	3.9998	81.962	1.154		
1	1.0000	3.4226	87.392	0.724	1	3
2	1.5883	3.0000	79.484	0.537	1	3
3	2.0000	2.6558	86.240	0.928	2	2
4	2.6558	2.0000	79.484	0.537	2	2
5	3.0000	1.5883	87.392	0.724	3	1
6	3.4226	1.0000	81.962	1.154	3	1
7	3.9998	0.0002				

The ground plan of the edge cables is indicated in Fig. 4.16. The connection of the cable net and the edge cable at the break points is ensured by the insertion of fictitious loads having the following magnitudes at the near points:

$X_1 = -4.060$

$X_2 = -0.215$

$X_3 = 0$

$X_4 = -0.382$

$X_5 = 4.060$

$X_6 = -0.215$

(X_1, X_3, X_5 are vertical joint loads, X_2, X_4, X_6 represent drawing-apart of cables.) Then, the final net joint heights in the x-directional cables are as follows (within the framed part the data is for the height of the real net joints; outside it, for the fictitious joints):

0.000	−1.750	−3 000	−3.750	−4.000	−3.750	−3.000	−1.750	0.000
1.750	−0.000	−1.271	−2.060	−2.409	−2.060	−1.271	−0.000	1.705
3.000	1.271	0.191	−0.834	−1.088	−0.834	0.191	1.271	3.000
3.750	2.275	0.834	−0.000	−0.272	−0.000	0.834	2.275	3.750
4.000	2.409	1.088	0.272	−0.000	0.272	1.088	2.409	4.000
3.750	2.275	0.834	−0.000	−0.272	−0.000	0.834	2.275	3.750
3.000	1.271	0.191	−0.834	−1.088	−0.834	0.191	1.271	3.000
1.750	−0.000	−1.271	−2.060	−2.409	−2.060	−1.271	−0.000	1.750
0.000	−1.750	−3.000	−3.750	−4.000	−3.750	−3.000	−1.750	0.000

The joint heights on the y-directional cables differ from the above data only by the values w. In the present case

$w_{1,3} = w_{1,5} = w_{3,1} = w_{3,7} = w_{5,1} = w_{5,7} = w_{7,3} = w_{7,5} = X_2 = -0.215$,

$w_{2,2} = w_{2,6} = w_{6,2} = w_{6,6} = X_4 = -0.382$ and the other w's are zero.

The heights of the break points of the cable are:

i	z_i
0	−4.000
1	−2.332
2	−1.339
3	−0.613
4	0.613
5	1.339
6	2.332
7	4.000

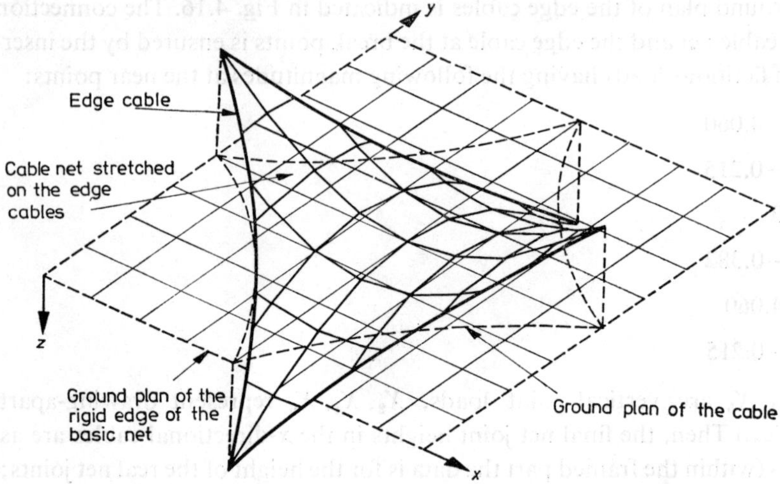

Fig. 4.16. Cable net stretched on four-edge cables

It can easily be checked that the break points of the edge cable do in fact lie on the nets. For instance:

$$z_1 = z_{1,3}+w_{1,3}+(z_{1,4}-z_{1,3}-w_{1,3})(y_1-y_{1,3})/b =$$
$$=-2.060-0.215+(-2.409+2.060+0.215)0.4226 = -2.332.$$

It can also be checked that Eq. (4.12; j, k) is satisfied on any joint of the real or fictitious nets. For instance:

On joint 1, 4: $\left(\frac{H_x}{a} = \frac{H_y}{b} = 10\right)$

$$10(4z_{1,4}-z_{0,4}-z_{1,3}-z_{1,5}-z_{2,4}-w_{1,3}-w_{1,5}) = p_{1,4} =$$
$$= 10(4(-2.409)+4.0+2.06+2.06+1.088+0.215+0.215) = 0.$$

On joint 1,3:

$$10(4z_{1,3}+2w_{1,3}-z_{0,3}-z_{1,2}-z_{1,4}-z_{2,3}) = p_{1,3} = X_1 =$$
$$= 10(4(-2.060)+2(-0.215)+3.75+1.271+2.409+0.834) = -4.060.$$

(In our example, the real net was unloaded, and only the fictitious joint loads act on the fictitious net.)

When comparing the height data of the basic net and the net supported by the edge cables, we can see what displacements were caused at the individual net points by introducing the edge cables. (This is greatest at point 1,4 — and at its symmetric counterpart — having the absolute value 0.159.) The net stretched on the edge cables is shown in Fig. 4.16.

4.5.2.2. Inner Closed Edge Cable

The ground plan of the closed edge cable can be uniquely constructed with a knowledge of one of its break points and the directions of its two sides emanating from this; for instance, the ground plan of the inner closed edge cable shown in Fig. 4.17 was constructed on the basis of point $0 \equiv \nu+1$ and directions $\hat{\mathbf{e}}_{0,1}$, $\hat{\mathbf{e}}_{\nu+1,\nu}$. In the case of a closed edge cable, the starting and the end points coincide, and the resultant of the horizontal components of the intersected cable forces of the supported rectangular net is automatically zero. The values of the horizontal cable force components belonging to the sides of the cable polygon can be determined by starting from the first break point (point 0), and progressing from side to side.

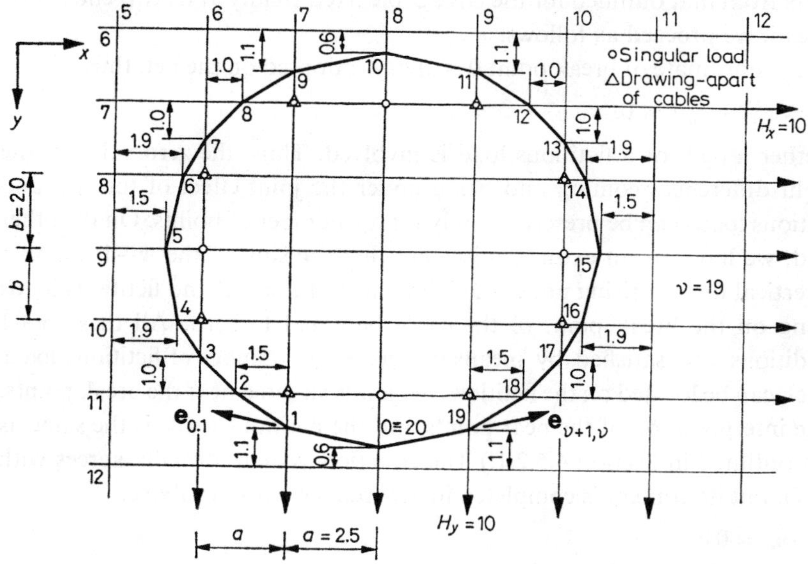

Fig. 4.17. Fictitious loads acting on a closed inner edge cable

If, for instance, the first break point — marked with 0 — is on the jth y-directional cable, then the sum of the horizontal components of the forces acting on point 0 is zero:

$$H_{0,1}\hat{\mathbf{e}}_{0,1}+H_{\nu+1,\nu}\hat{\mathbf{e}}_{\nu+1,\nu}\pm H_{y,j}\hat{\mathbf{e}}_y = 0,$$

where the sign of the third term must be adopted in such a manner that the corresponding cable force should point outward from the part of the net limited by the edge cable. From this equilibrium equation and with knowledge of $\hat{\mathbf{e}}_{0,1}$, $\hat{\mathbf{e}}_{\nu+1,\nu}$, H_{yj} and $\hat{\mathbf{e}}_y$, $H_{0,1}$ and $H_{\nu+1}$, can be calculated (on condition that $\hat{\mathbf{e}}_{0,1}$ and $\hat{\mathbf{e}}_{\nu+1,\nu}$ do not lie on the same straight line):

$$\begin{bmatrix} H_{0,1} \\ H_{\nu+1,\nu} \end{bmatrix} \pm [\hat{\mathbf{e}}_{0,1}\hat{\mathbf{e}}_{\nu+1,\nu}]^{-1} H_{yj}\hat{\mathbf{e}}_y = 0.$$

The direction of the second polygon side and the value of the horizontal cable force component belonging to it can be determined from the equilibrium equation for break point 1. Assuming that this break point is on the $(j-1)$th y-directional cable,

$$H_{1,0}\hat{\mathbf{e}}_{1,0} + H_{1,2}\hat{\mathbf{e}}_{1,2} \pm H_{y,j-1}\hat{\mathbf{e}}_y = 0,$$

from which $H_{1,2}\hat{\mathbf{e}}_{1,2}$ can be determined, and with knowledge of $\hat{\mathbf{e}}_{1,2}$, the position of the next break point — marked by 2 — can be calculated.

The determination of the spatial position of the inner closed edge cable differs from that outlined for the edge cable fixed rigidly at its two ends only. Instead, we proceed as follows.

The z_0 coordinate of break point 0 is always adopted on the net, thus

$$\Delta z_0 = z_0^{(h)} - z_0^{(k)} = 0,$$

whether a real, or a fictitious load is involved. Thus, the zero value of the height differences coming into being under the joint effect of the real and fictitious loads can be prescribed only for a ν number of points. On the other hand, we have, as an extra requirement that the sum of the $\nu+1$ number of vertical forces arising under the joint effect of the real and fictitious loads acting on the break point of the cable be equal to zero. All these $\nu+1$ conditions are satisfied by means of the $\nu+1$ number of fictitious loads which can be located on the fictitious net joints situated near the break points. (The interpretation of the near points and the fictitious loads is the same as that outlined in Section 4.5.2.1.) The equation system formally agrees with (4.17), but its content is completed in accordance with the above.

$$\mathbf{A}\mathbf{x} + \mathbf{a}_0 = 0$$

$$\mathbf{x} = \begin{bmatrix} X_1 \\ X_2 \\ \vdots \\ X_{\nu+1} \end{bmatrix} \quad \mathbf{a}_0 = \begin{bmatrix} a_{0,1} \\ a_{0,2} \\ \vdots \\ a_{0,\nu+1} \end{bmatrix}$$

$$\mathbf{A} = [a_{i,\mu}] \quad (i, \mu = 1, 2, \ldots, \nu+1), \tag{4.19}$$

$$a_{0,\sigma} = \begin{cases} = \Delta z_\sigma, & \text{if } \sigma = 1, 2, \ldots, \nu, \\ = \sum_{i=0}^{\nu} P_i, & \text{if } \sigma = \nu+1, \end{cases}$$

$$a_{\sigma,\mu} = \begin{cases} = \Delta z_{\sigma,\mu}, & \text{if } \sigma = 1, 2, \ldots, \nu, \\ = \sum_{i=0}^{\nu} P_{i,\mu}, & \text{if } \sigma = \nu+1, \end{cases}$$

here Δz_σ is the difference in height

$$\Delta z_\sigma = z_\sigma^{(h)} - z_\sigma^{(k)}$$

corresponding to the σth break point in the case of the basic load, while $z_{\sigma,\mu}$ is the height corresponding to the σth break point in the case of the μth unit fictitious load. Moreover, P_i is the sum of the vertical components of the cable forces acting in the ith intersected cable, pointing from the edge towards the real part of the net and arising under the basic load and the vertical load acting there; similarly for $P_{i,\mu}$ which arises under the μth unit fictitious load and the vertical load acting there. (In the case of a fictitious load, $P_{i,\mu}$ contains only the vertical component of the cable force!)
It is not difficult to see that the sum of the vertical components of the cable forces acting on the closed edge cable from the direction of the real part of the net, $\sum_{i=0}^{\nu} P_i$, differs only in sign from the sum of the vertical forces $P_{j,k}$ acting on the joints of the so-called fictitious part of the net within the closed edge. Thus, if the μth unit fictitious load acts on a "near" net point selected for the first time, then this fictitious load will be a unit vertical load, and thus

$$a_{\nu+1,\mu} = \sum_{i=0}^{\nu} P_{i,\mu} = -1.$$

If the μth fictitious load acts on such a "near" net point which occurs for the second time as a near point, then the fictitious load will be a $w_{j,k}=1$ unit drawing-apart of cables. According to (4.12), this unit fictitious load is equivalent to three vertical forces

$$p_{j,k-1} = \frac{H_{yj}}{b}, \quad p_{j,k} = -\frac{2H_{yj}}{b}, \quad p_{j,k+1} = \frac{H_{yj}}{b},$$

of which only the first two, or only the last two, act on the joint of the fictitious part of the net within the closed cable edge; in this case:

$$a_{\nu+1,\mu} = \sum_{i=0}^{\nu} P_{i,\mu} = \frac{H_{yj}}{b}.$$

Example 4.7. Let us consider the net shown in Fig. 4.7 as a basic net, and let $H_x = H_y = 10$, $a = 2,5$, $b = 2,0$, $z_{8,5} = 10$ and $z_{0,0} = z_{0,16} = 3.2$. Let us support a part of the net by an inner closed edge cable the starting point of which (marked by 0) has the coordinates

$$x_0 = 20, \quad y_0 = 23.4.$$

Let the directions of the two polygon sides joining point 0 be

$$\hat{\mathbf{e}}_{0,1} = \begin{bmatrix} -0.980\,581 \\ -0.196\,116 \end{bmatrix}; \quad \hat{\mathbf{e}}_{v+1,v} = \begin{bmatrix} 0.980\,581 \\ -0.196\,116 \end{bmatrix}.$$

(See Fig. 4.17.)
In the present case,

$$H_{0,1}\mathbf{e}_{0,1} + H_{v+1,v}\hat{\mathbf{e}}_{v+1,v} + H_y\hat{\mathbf{e}}_y = 0,$$

that is,

$$\begin{bmatrix} H_{0,1} \\ H_{v+1,v} \end{bmatrix} = [\mathbf{e}_{0,1}\mathbf{e}_{v+1,v}]^{-1} H_y\hat{\mathbf{e}}_y = \begin{bmatrix} 25.495\,12 \\ 25.495\,12 \end{bmatrix}.$$

Concerning break point 1

$$H_{1,0}\hat{\mathbf{e}}_{1,0} + H_{1,2}\hat{\mathbf{e}}_{1,2} + H_y\hat{\mathbf{e}}_y = 0,$$

thus

$$H_{1,2}\mathbf{e}_{1,2} = H_{0,1}\hat{\mathbf{e}}_{0,1} - H_y\hat{\mathbf{e}}_y =$$

$$= \begin{bmatrix} -25.0 \\ -\ 5.0 \end{bmatrix} - \begin{bmatrix} 0 \\ 10.0 \end{bmatrix} = \begin{bmatrix} -25.0 \\ -15.0 \end{bmatrix} = 29.154\,76 \begin{bmatrix} -0.857\,493 \\ -0.514\,496 \end{bmatrix}.$$

With a knowledge of the direction $\hat{\mathbf{e}}_{1,2}$, the location of break point 2 and the length of the ground plan of the polygon side 1, 2 are determined: $d_{1,2}=1.749\,286$. Thus we progress from point to point until all values $H_{i,i+1}$, $d_{i,i+1}$ belonging to every cable in the polygon are calculated. After this, knowing the net heights $z_i(h)$ belonging to the vertical of the break points, the vertical components P_i of the cable forces (these being in the vertical of the break points) as well as with the aid of equation system (4.16, z), we can determine the coordinates $z_i^{(k)}$ of the break points of the cable; then the height differences $\Delta z_i = z_i^{(h)} - z_i^{(k)}$ are formed:

i	P_i	$z_i^{(h)}$	$z_i^{(k)}$	Δz_i	
0	−3.108	1.872 60	1.872 60	0	
1	−2.894	2.068 45	2.027 25	0.041 20	
2	0.216	2.361 40	2.293 74	0.067 66	
3	−2.379	2.621 00	2.575 44	0.045 56	
4	−0.044	2.856 36	2.839 66	0.016 70	
5	−0.468	3.279 20	3.371 62	−0.092 42	$\sum_{i=0}^{19} P_i = 0$
6	−0.996	3.730 24	3.941 02	−0.210 78	
7	2.261	4.016 00	4.265 56	−0.249 56	
8	−2.063	4.448 40	4.655 79	−0.207 39	
9	4.061	5.123 40	5.130 84	−0.007 44	
10	7.720	6.217 10	5.516 59	0.070 51	

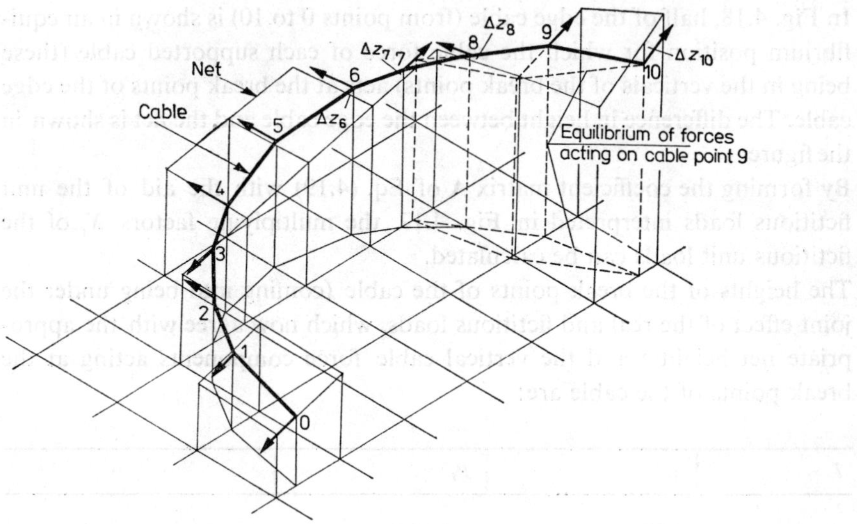

Fig. 4.18. Equilibrium of the break points of the inner edge cable

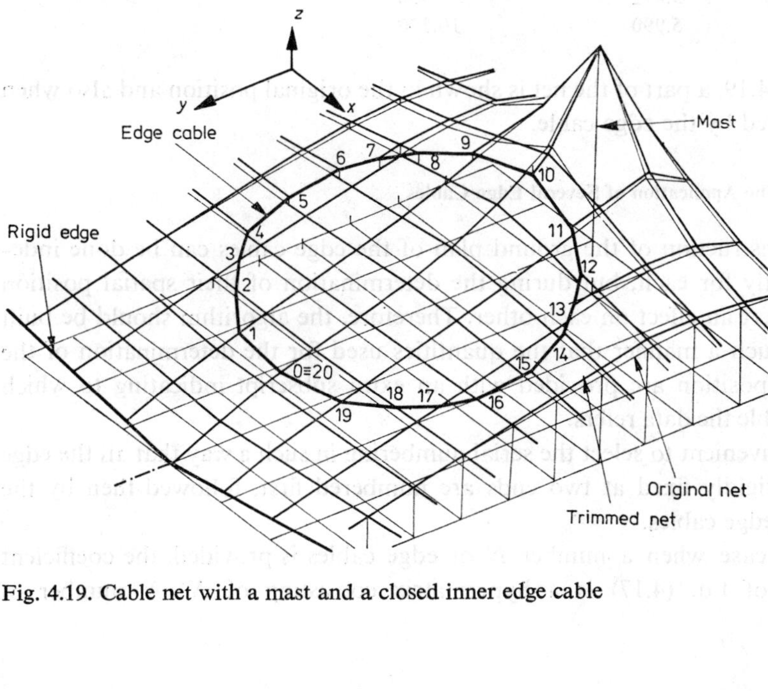

Fig. 4.19. Cable net with a mast and a closed inner edge cable

In Fig. 4.18, half of the edge cable (from points 0 to 10) is shown in an equilibrium position for which the cable force of each supported cable (these being in the verticals of the break points) acts at the break points of the edge cable. The difference in height between the edge cable and the net is shown in the figure.

By forming the coefficient matrix **A** of Eq. (4.19) with the aid of the unit fictitious loads interpreted in Fig. 4.17, the multiplying factors X_i of the fictitious unit loads can be calculated.

The heights of the break points of the cable (coming into being under the joint effect of the real and fictitious loads, which now agree with the appropriate net heights) and the vertical cable force components acting at the break points of the cable are:

i	z_i	P_i	
0	1.918	−3.335	
1	2.084	−2.898	
2	2.358	0.085	
3	2.655	−2.891	
4	2.950	0.088	
5	3.532	−0.668	$\sum_{i=0}^{19} P_i = 0.001$
6	4.167	−1.533	
7	4.548	2.894	
8	4.987	−1.494	
9	5.472	2.900	
10	5.990	10.370	

In Fig. 4.19, a part of the net is shown in the original position and also when supported by the edge cable.

4.5.2.3 The Application of Several Edge Cables

The construction of the ground plan of the edge cables can be done independently for each, but during the determination of their spatial position they exert an effect on each other. Therefore, the algorithm should be built up in such a manner that the quantities used for the determination of the spatial position are provided with an extra subscript indicating to which edge cable the data refers.

It is convenient to select the serial numbering in such a way that all the edge cables rigidly fixed at two ends are numbered first, followed then by the closed edge cables.

In the case when a number N of edge cables is provided, the coefficient matrix of Eq. (4.17) is a hypermatrix consisting of $N \times N$ number of blocks. If, of the N number of edge cables, a number M are fixed at the two ends, then the number of the rows ($q_{J,K}$) and columns ($p_{J,K}$) of block $\mathbf{A}_{J,K}$ is:

$$q_{J,K} = \begin{cases} \nu_J, & \text{if } J \leq M \\ \nu_{J+1}, & \text{if } J > M, \end{cases}$$

$$(J, K = 1, 2, \ldots, N)$$

$$p_{J,K} = \begin{cases} \nu_K, & \text{if } K \geq M \\ \nu_K + 1, & \text{if } K > M, \end{cases}$$

where ν now denotes the number of the intermediate break points of the edge cable (not including the starting point 0 and the end point $\nu+1$ among them).

In general, the τth element of the σth row of the block

$$\mathbf{A}_{J,K} = [a_{J,K,\sigma,\tau}]; \quad \begin{pmatrix} \sigma = 1, 2, \ldots, q_{J,K} \\ \tau = 1, 2, \ldots, p_{J,K} \end{pmatrix}$$

stands for the difference in height between the net and the edge cable above the σth break point of the Jth edge cable, in the case of the τth fictitious load of the Kth edge cable. An exception to this is the (ν_J+1)th row in the case $J > M$. Then,

$$a_{J,K;\nu_J+1,\tau} = \begin{cases} 0, & \text{if } J \neq K\ * \\ -1, & \text{if } J = K \text{ and the near net joint of the } (\tau-1)\text{th cable break point occurs for the first time;} \\ \dfrac{H_{y,j_{\tau-1}}}{b} & \text{if } J = K \text{ and the near net joint of the } (\tau-1)\text{th cable break point occurs for the second time.} \end{cases}$$

Vector $\mathbf{a}_0$ of Eq. (4.17) will also be a hypervector, whose Jth vector can be determined by formula (4.18) in the case when $J \leq M$, and by formula (4.19) in the case when $J > M$.

If the net is lifted by masts into the prescribed height position at selected points, then this condition must be satisfied, in the case of the basic load, by the same number of supplementary solutions as the number of masts. In loading cases corresponding to the fictitious loads, zero height must be ensured at the places of the masts by making use of the supplementary solutions.

* Namely, in such a case the fictitious part of the net within the Jth closed edge cable is unloaded.

4.6. Other Net Types Constructed with the Aid of a Rectangular Cable Net

The determination of the spatial position of the rectangular cable net stretched on the rigid edge or on edge cables, possibly supported by masts at some of its points, is a relatively simple task as discussed in Section 4.5. For this purpose, algorithms suitable for programming can be constructed. The solution of Eq. (4.13), forming the basis of the calculation work, requires only the multiplication of matrices of mth and nth order, respectively, even in the case of a net consisting of $m+n$ number of cables (thus, mn number of inner joints), and hence the edge fitting of large-sized nets, can also be calculated by computers having a relatively small capacity.

The rectangular cable net, because of its simple calculation, is frequently accepted as a final net too, especially if one disregards the fact that the meshes having a rectangular shape in ground plan are rhomboidal in space. This does not present any particular problem if the net is stretched on an edge that can be regarded as rigid (this is always assumed during the determination of the erection shape). However, if the net is supported by masts at its intermediate points, then the cable forces may considerably increase near the masts, and in the case of the rectangular net, the strengthening or "densification" of the cables presents a new problem. In such a case, it is desirable to regard the rectangular net only as a means of calculation and, after determining the lines of principal curvature of a surface definable by means of it, to construct a new net with cables running along these lines. The net constructed in this manner is called the net of principal curvatures. The main load-bearing cables of this net form converging lines in the vicinity of the mast; thus they follow the actual internal forces of the net more closely, and the perpendicular principal curvature lines constitute a system of concentric lines near the mast. The cables running along these lines are of great importance in the stiffening of the net.

The requirement may also arise that the net should be constructed with cables lying on the surface defined by the rectangular net, led along the geodetic lines. In the erection state only small shear forces arise between the intersecting cables of such a net.

Of course, in the loaded state (i.e. during the variation of the load) a considerable shear force may arise between the intersecting cables.

In what follows, the mode of construction of the net of principal curvatures will be discussed in more detail, and only brief reference will be made to the geodetic net. Prior to this, a method for constructing an interpolation surface, allocated to the rectangular net, suitable for the purpose of a — mathematically not exact — calculation will be discussed. The interpolation surface serves exclusively for interpolating surface lines with the aid of the differential–geometrical quantities to be defined on it. The very precise determination of these lines has no particular significance because, irrespective of whether a net of principal curvatures or a geodetic net is constructed, the cables of the net will not be smooth surface lines, but continuous spatial lines having break points at a finite number of points on each surface curve.

4.6.1. Interpolation Surface

The rectangular cable net fitted to the prescribed edge, possibly supported by masts at some of its intermediate points, is considered as given. The coordinates of every net joint and edge point, the cable forces, and the cable forces of the edge cables — if the latter are part of the system — are known.
Within the ground plan of the net, we define the subregion allocated to the inner net joint j, k by the condition that

$$x_j - \frac{a}{2} < x < x_j + \frac{a}{2},$$

$$y_k - \frac{b}{2} < y < y_k + \frac{b}{2}.$$

In the neighbourhood of the edge there are such areas — limited by the median straight lines of the cable sections and the edge — which do not be-

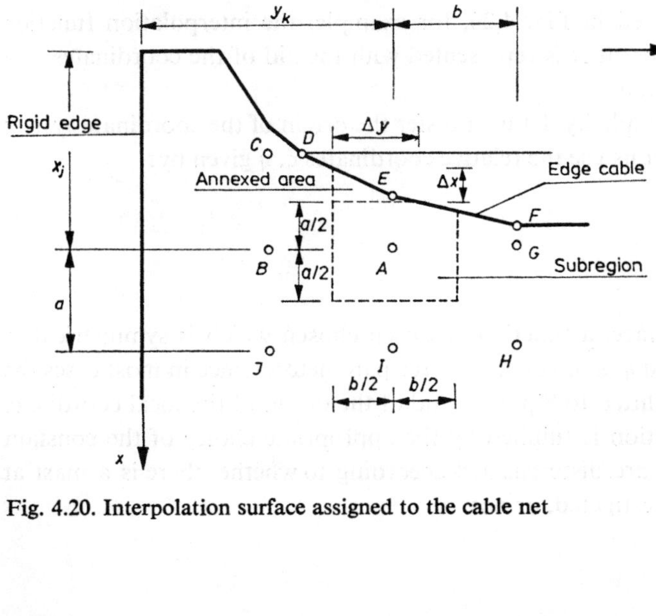

Fig. 4.20. Interpolation surface assigned to the cable net

long to either of the subregions as defined above. These areas are always annexed to that subregion which they adjoin. An area adjoining several subregions belongs to that subregion for which the common side is parallel to the x-axis, if

$$\frac{\Delta x}{a} \geqslant \frac{\Delta y}{b};$$

otherwise it is annexed to the subregion having a common boundary with the y-axis.
(Here Δx indicates the length of the boundary of the investigated area parallel to the x-axis, and Δy that of the boundary parallel to the y-axis). The subregion of some surface elements belonging to a few joints — lying near the boundary — are shown in Fig. 4.20.
The interpolation functions applied below are continuous throughout the subregion — with the exception, at most, of a single point — , and can be differentiated at least twice according to the variables x, y. The totality of the interpolation functions belonging to the entire net has a discontinuity where the subregions are continuous. The interpolation function belonging to the subregions is represented by those joints of the net as well as those points of intersection of the edge with the net the coordinates of which satisfy the following conditions:

$$x_j - a \leqslant x \leqslant x_j + a,$$

$$y_k - b \leqslant y \leqslant y_k + b.$$

In the case depicted in Fig. 4.20, for example, the interpolation function belonging to net point A is represented with the aid of the coordinates z of points A, B, C, ..., J.
For the sake of simplicity, let us transfer the origin of the coordinate system to point A, and let us use the relative coordinates ξ, η given by:

$$\xi = x - x_A,$$

$$\eta = y - y_A.$$

To define the surface, a function has been chosen which is symmetrical in the variables ξ and η, and contains 8 free parameters (since in most cases the surface has to be fitted to 9 points, and at the origin of the local coordinate system, the condition is fulfilled by the appropriate choice of the constant term). Two cases are distinguished according to whether there is a mast at the inner joint investigated, or not.

If the inner point A under investigation is not supported by a mast, then the surface is sought in the form

$$F(\xi, \eta) = z_A + c_{A1}\xi + c_{A2}\eta + c_{A3}\xi^2 + c_{A4}\xi\eta + c_{A5}\eta^2 + c_{A6}\xi^3 + c_{A7}\xi^2\eta + c_{A8}\xi\eta^2 + c_{A9}\eta^3. \tag{4.20}$$

This surface assumes the prescribed height at point A. We endeavour to fulfil the condition of fitting the points by the proper selection of coefficient $c_{A\nu}$ $(\nu=1, 2, \ldots, 8)$. If the number of fitting points outside point A is n_A, then the equations

$$z_i - z_A = [\xi\eta\ \xi^2\ \xi\eta\ \eta^2\ \xi^2\eta\ \xi\eta^2\ \xi^2\eta^2]_i \mathbf{c}_A = \mathbf{x}_i^* \mathbf{c}_A \quad (i = 1, 2, \ldots, n_A) \tag{4.21}$$

have to be satisfied. If $n_A < 8$, then the system of equations (4.21) is indeterminate; therefore, in such a case the coefficients of the terms higher than the second degree are made equal to zero. If the equation system is over-determinate, then the solution is sought with the aid of the least-square error procedure. Thus

$$\mathbf{c}_A = (\mathbf{X}_A^* \mathbf{X}_A)^{-1} \mathbf{X}_A^* \Delta \mathbf{z}_A, \tag{4.22}$$

where

$$\mathbf{X}_A = \begin{bmatrix} \mathbf{x}_2^* \\ \mathbf{x}_3^* \\ \vdots \\ \mathbf{x}_{n_A}^* \end{bmatrix}; \quad \Delta \mathbf{z}_A = \begin{bmatrix} z_2 - z_1 \\ z_3 - z_1 \\ \vdots \\ z_{n_A} - z_1 \end{bmatrix}.$$

When $|\mathbf{X}_A| \neq 0$, the expression following from (4.22) is also used:

$$\mathbf{c}_A = \mathbf{X}_A^{-1} \Delta \mathbf{z}_A.$$

Let us now examine how the interpolation function has to be modified if the inner point investigated (i.e. point A, this being the point of intersection of the ith y-directional and the jth x-directional cable) is supported by a mast. There is usually no edge in the vicinity of an inner mast, so that hereinafter we shall presume that the four cable sections starting from the mast join the inner points of the cable net. Then, the function $F(\xi, \eta)$ (4.20) is completed by the additional term

$$\psi \ln\left(\sqrt{\left(\frac{\xi}{a}\right)^2 + \left(\frac{\eta}{b}\right)^2} + 1\right), \tag{4.23}$$

where

$$\psi = \frac{z_{i,j+1} + z_{i,j-1} + z_{i+1,j} + z_{i-1,j} - 4z_{i,j}}{4 \ln 2}.$$

Formula (4.23) was derived from the condition that it should give zero height at point A, while at the neighbouring four inner points it should yield equal values, i.e. the average of their deviation from the height at point A; also, its shape should approximate the surface in the vicinity of the mast. The parameters of the function F are again calculated by Eq. (4.22) but the meaning of $\Delta \mathbf{z}_A$ is now different:

$$\mathbf{z}_A = \begin{bmatrix} z_1 - \psi \ln(\varrho_1 + 1) - z_A \\ z_2 - \psi \ln(\varrho_2 + 1) - z_A \\ \vdots \\ z_{n_A} - \psi \ln(\varrho_{n_A} + 1) - z_A \end{bmatrix},$$

where

$$\varrho_i = \sqrt{\left(\frac{\xi_i}{a}\right)^2 + \left(\frac{\eta_i}{b}\right)^2}.$$

4.6.2. The Construction of the Net of Principal Curvatures

The geometrical conditions of the "net surface" can be investigated with the aid of the interpolation functions belonging to the subregions allocated to the net joints. Within these subregions it is possible to calculate, point by point, the derivatives of the function $z = F(\xi, \eta)$;

$$p = \frac{\partial F}{\partial \xi}; \quad q = \frac{\partial F}{\partial \eta};$$

$$r = \frac{\partial^2 F}{\partial \xi^2}; \quad s = \frac{\partial^2 F}{\partial \xi \partial \eta}; \quad t = \frac{\partial^2 F}{\partial \eta^2}.$$

$$\begin{bmatrix} p \\ q \\ r \\ s \\ t \end{bmatrix} = \mathbf{Q}(\xi, \eta)\mathbf{c}_A + \psi \begin{bmatrix} \dfrac{\xi}{a^2(\varrho^2+\varrho)} \\ \dfrac{\eta}{b^2(\varrho^2+\varrho)} \\ \dfrac{1}{a^2(\varrho^2+\varrho)} - \dfrac{\xi^2(2\varrho+1)}{a^4\varrho(\varrho^2+\varrho)^2} \\ \dfrac{-\xi\eta(2\varrho+1)}{a^2b^2\varrho(\varrho^2+\varrho)^2} \\ \dfrac{1}{b^2(\varrho^2+\varrho)} - \dfrac{\eta^2(2\varrho+1)}{b^4\varrho(\varrho^2+\varrho)^2} \end{bmatrix}$$

where the second term of the right side occurs only in the case when point A is supported by a mast, and

$$\mathbf{Q} = \begin{bmatrix} 1 & 0 & 2\xi & \eta & 0 & 2\xi\eta & \eta^2 & 2\xi\eta^2 \\ 0 & 1 & 0 & \xi & 2\eta & \xi^2 & 2\xi\eta & 2\xi^2\eta \\ 0 & 0 & 2 & 0 & 0 & 2\eta & 0 & 2\eta^2 \\ 0 & 0 & 0 & 1 & 0 & 2\xi & 2\eta & 4\xi\eta \\ 0 & 0 & 0 & 0 & 2 & 0 & 2\xi & 2\xi^2 \end{bmatrix}.$$

At the point of location of the mast the derivatives are not defined.
The tangent unit vectors of the principal curvatures

$$\mathbf{e}_1 = \begin{bmatrix} \alpha_1 \\ \beta_1 \\ \gamma_1 \end{bmatrix}; \quad \mathbf{e}_2 \begin{bmatrix} \alpha_2 \\ \beta_2 \\ \gamma_2 \end{bmatrix}$$

are determined by the expressions

$$\left.\begin{aligned} \alpha_i &= \frac{1}{\sqrt{1+m_i^2+(p+qm_i)^2}} \\ \beta_i &= m_i\alpha_i \\ \gamma_i &= p\alpha_i+q\beta_i \end{aligned}\right\} \quad (i = 1,2),$$

where m_1 and m_2 are the roots of the equation

$$h_2m^2+h_1m+h_0 = 0$$

and

$$h_2 = pqt-(1+q^2)s,$$

$$h_1 = (1+p^2)t-(1+q^2)r,$$

$$h_0 = (1+p^2)s-pqr.$$

If $h_2=0$, but $h_1 \neq 0$, then $m_1 = -\dfrac{h_0}{h_1}$ and

$$\alpha_2 = 0,$$

$$\beta_2 = \frac{1}{\sqrt{1+q^2}},$$

$$\gamma_2 = q\beta_2.$$

The set of lines of principal curvature of the "net surface" can be constructed with arbitrary density. The construction is done in two steps. In the first step, the projection on the x, y plane, of a line of curvature passing through a selected point is determined; in the second step, the lines of the ground plan are projected onto the "net surface". For the construction of the ground

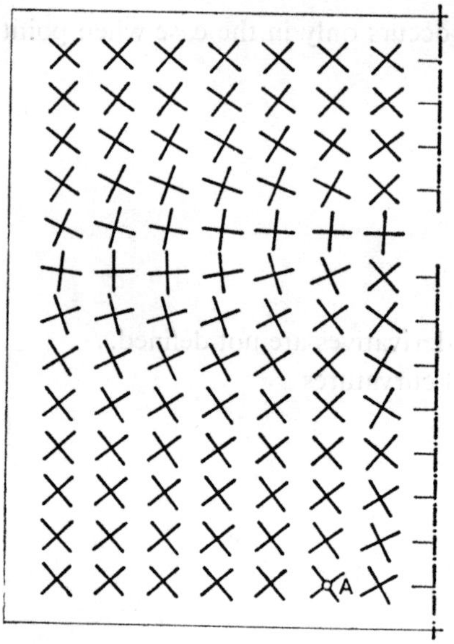

Fig. 4.21. Direction field of principal curvatures

plan projection of the principal curvature lines the interpolation functions belonging to the subregions are used. The discontinuity of the totality of the function at the boundary of the subregions does not cause any difficulties. The ground plan of the principal curvature lines can be constructed graphically (manually), or with the aid of a computer. For the graphical construction we produce, for each subregion and with the required density, the field of tangent-directions given by the inclinations

$$m_1 = \frac{\alpha_1}{\beta_1}, \quad m_2 = \frac{\alpha_2}{\beta_2}, \quad \text{i.e., directions } \hat{\mathbf{e}}_1, \hat{\mathbf{e}}_2.$$

(In Fig. 4.21 the directions belonging to net joints of the direction field of the net shown in Fig. 4.8 are given). Once the direction field of the ground plan has been established, the curve is constructed by numerical integration, starting from some point on the ground plan, and taking into consideration the fact that the curves, in general, do not pass smoothly across the boundary of the subregions.

For the construction of the principal curvature lines, the first line can be started from an arbitrary point of the direction field (the first line started from point A of the direction field shown in Fig. 4.21 is presented in Fig. 4.22). Once this first line has been constructed, we mark the points of intersection of the other set of lines (points $B, C, \ldots, M$ of Fig. 4.22). These points can be marked in such a manner that they are located with equal spacing when projected back to the "net surface", or on the basis of other

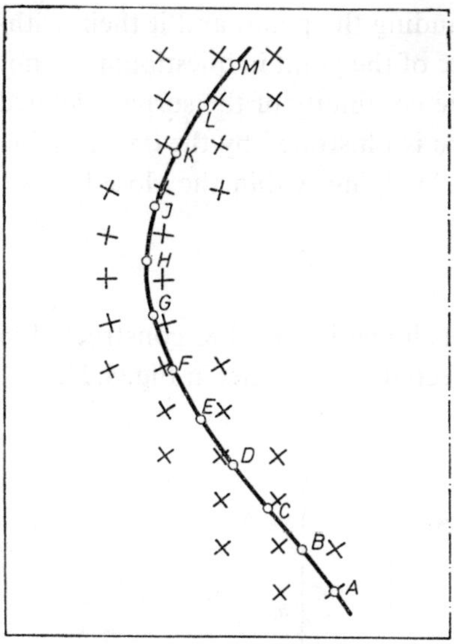

Fig. 4.22. Cable fitting the direction field of principal curvatures

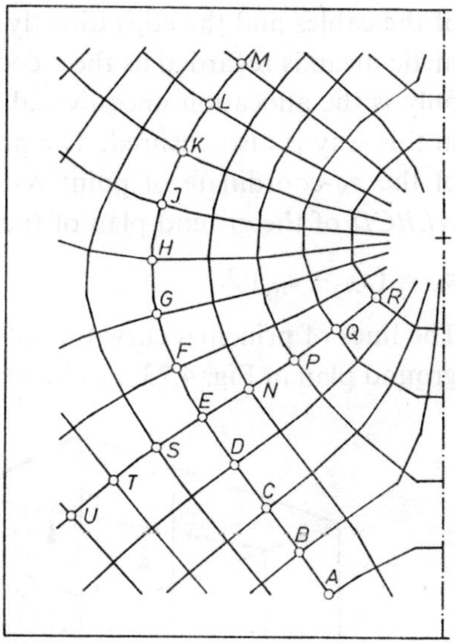

Fig. 4.23. Cable net of principal curvatures

considerations. (The points of Fig. 4.22 are located at equal distances, these being measured on the surface.)

The second line of principal curvature (lines *E–R* and *E–U* in Fig. 4.23) — perpendicular to the first one — is started from some point on the first principal curvature line (e.g. from point *E* in Fig. 4.22). Then, the points of intersection of the other principal curvature lines are marked out on this line also, on the basis of some consideration. These other principal curvature lines are constructed by starting from the points of intersection marked out on the first two lines, with the required density.

When the principal curvature lines have been produced on the ground plan with the desired density, we can project them back to the "net surface" — if this is required. Namely, in general, we only need the *z*-coordinates of the points of intersection of the principal curvature lines, because the cables of the so-called "net cable of principal curvature" are spatial lines connecting the points of intersection of the principal curvature lines as break points; we cannot speak of orthogonality for these although the curves on which the break points lie constitute an orthogonal set of curves. Thus, it is sufficient to present a method for the projection of the points of the ground plan that makes possible the construction of continuous surface lines. The interpolation functions used in the foregoing are not suitable for this purpose. If, however, we compute two values for the height of the point to be projected by linear interpolation in the *x* and *y* directions, respectively, from the data of the cables and the edge directly surrounding the point, and if their arithmetic mean is regarded as the z-coordinate of the point in question, then not only is the allocation unequivocal, but the continuity of the surface defined in this way is also ensured. The procedure is illustrated by the construction of the z_N-coordinate of point N (Fig. 4.24) lying within the closed figure $ALBCD$ of the ground plan of the net:

$$z_N = (z_{N_x} + z_{N_y})/2.$$

The lines of principal curvature of the net shown in Fig. 4.8, constructed in ground plan in Fig. 4.23, can be seen projected onto the net in Fig. 4.25.

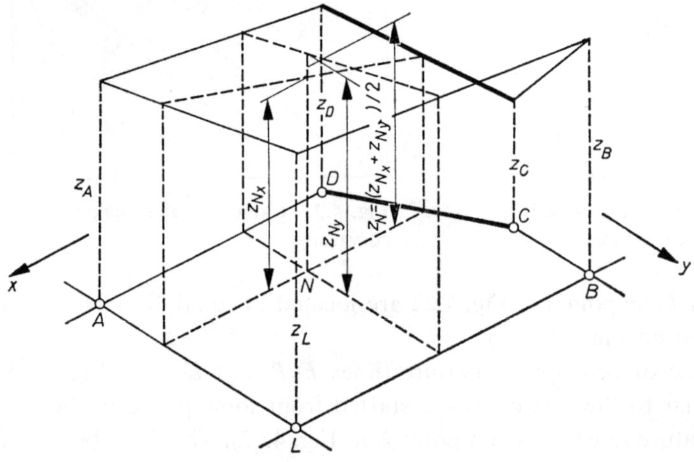

Fig. 4.24. Interpolation of nodal heights of a cable net of principal curvatures

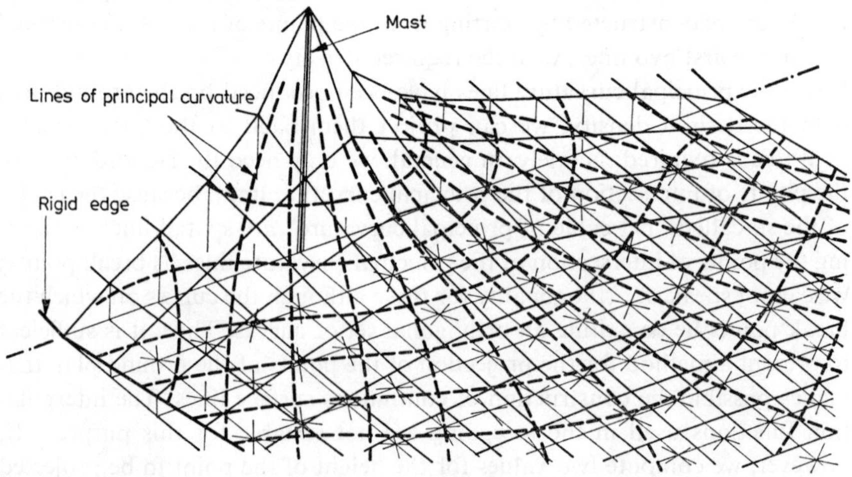

Fig. 4.25. Cable net of principal curvatures stretched on a mast

The approximate values of the cable forces in the net of principal curvatures can be determined with the aid of the relationship between the cable force intensity in the net of principal curvatures and that of the rectangular net, both of infinitesimal density.

The point on the net with x, y, z coordinates and the geometrical data for the neighbourhood of this point are given. We want to establish the cable force intensity that corresponds to the rectangular cable net and to the cable net of principal curvatures in the infinitesimally small neighbourhood lying on the tangent-plane of the point in question.

The intensity of the horizontal components of the cable forces in the rectangular cable net is given by the values $h_x = H_x/b$, $h_y = H_y/a$. First the intensity of these cable forces in the tangent-plane must be determined. For this purpose, let us obtain the unit vectors $\mathbf{e}_{(x)}$, $\mathbf{e}_{(y)}$ of the line of intersection of the tangent-plane having unit normal vector $\mathbf{n}$ and the planes having normal vectors $\mathbf{e}_y$, $\mathbf{e}_x$ (that is, the coordinate planes x, z and y, z or the planes parallel to them) as well as the vectors $\mathbf{e}_{n(x)}$, $\mathbf{e}_{n(y)}$, in the tangent-plane, which are perpendicular to these (Fig. 4.26a):

$$\mathbf{n} = \frac{1}{\sqrt{1+p^2+q^2}} \begin{bmatrix} -p \\ -q \\ 1 \end{bmatrix},$$

$$\mathbf{e}_{(x)} = \frac{1}{\sqrt{1+p^2}} \begin{bmatrix} 1 \\ 0 \\ p \end{bmatrix}, \quad \mathbf{e}_{n(x)} = \frac{1}{\sqrt{(1+p^2)(1+p^2+q^2)}} \begin{bmatrix} -pq \\ 1+p^2 \\ q \end{bmatrix},$$

$$\mathbf{e}_{(y)} = \frac{1}{\sqrt{1+q^2}} \begin{bmatrix} 0 \\ 1 \\ q \end{bmatrix}, \quad \mathbf{e}_{n(y)} = \frac{1}{\sqrt{(1+q^2)(1+p^2+q^2)}} \begin{bmatrix} 1+q^2 \\ -pq \\ p \end{bmatrix},$$

$$\mathbf{n}^*\mathbf{e}_{(x)} = \mathbf{n}^*\mathbf{e}_{(y)} = \mathbf{n}^*\mathbf{e}_{n(x)} = \mathbf{n}^*\mathbf{e}_{n(y)} = \mathbf{e}^*_{(x)}\mathbf{e}_{n(x)} = \mathbf{e}^*_{(y)}\mathbf{e}_{n(y)} = 0.$$

The value of an x-directional horizontal cable force component belonging to a band of width Δy is

$$h_x \Delta y,$$

and the value of the corresponding cable force is

$$\frac{h_x \Delta y}{\mathbf{e}^*_x \mathbf{e}_{n(x)}}.$$

However, this cable force is distributed, in the tangent plane, throughout a band of width

$$\frac{\Delta y}{\mathbf{e}^*_y \mathbf{e}_{n(x)}}.$$

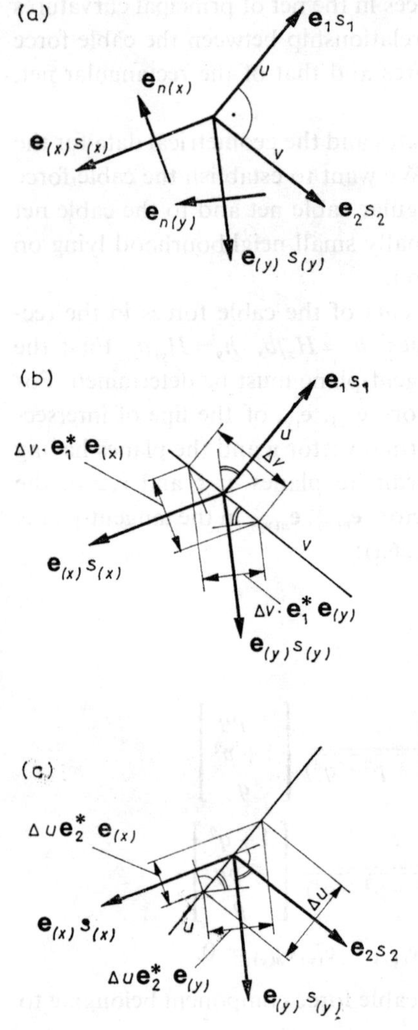

Fig. 4.26. Relations between cable forces of the rectangular net and of that of principal curvatures

Thus the cable force intensity corresponding to h_x in the tangent plane is

$$s_{(x)} = \frac{h_x \Delta y}{\mathbf{e}_x^* \mathbf{e}_{(x)}} : \frac{\Delta y}{\mathbf{e}_y^* \mathbf{e}_{n(x)}} = h_x \frac{\mathbf{e}_y^* \mathbf{e}_{n(x)}}{\mathbf{e}_x^* \mathbf{e}_{(x)}},$$

or, with the usual notation:

$$s_{(x)} = h_x \frac{1+p^2}{\sqrt{1+p^2+q^2}}. \tag{4.24}$$

In a similar manner we obtain

$$s_{(y)} = h_y \frac{\mathbf{e}_x^* \mathbf{e}_{n(y)}}{\mathbf{e}_y^* \mathbf{e}_{(y)}},$$

or

$$s_{(y)} = \frac{1+q^2}{\sqrt{1+p^2+q^2}}. \tag{4.25}$$

The cable force intensities s_1, s_2 of the principal curvatures are determined by the condition that $s_1 \Delta v$ should be equal to the sum of the u-directional cable force components, i.e. $s_{(x)}$ and $s_{(y)}$ belonging to Δv,

$$s_1 \Delta v = s_{(x)} \Delta v \mathbf{e}_1^* \mathbf{e}_{(x)} \mathbf{e}_1^* \mathbf{e}_{(x)} + s_{(y)} \Delta v \mathbf{e}_1^* \mathbf{e}_{(y)} \mathbf{e}_1^* \mathbf{e}_{(y)},$$

or, using the previous notation,

$$s_1 = h_x \frac{1+p^2}{\sqrt{1+p^2+q^2}} \left(\frac{\alpha_1 + p\gamma_1}{\sqrt{1+p^2}} \right)^2 + h_y \frac{1+q^2}{\sqrt{1+p^2+q^2}} \left(\frac{\beta_1 + q\gamma_1}{\sqrt{1+q^2}} \right)^2.$$

From the expression for the unit vector normal to the surface:

$$\mathbf{n} = \begin{bmatrix} \lambda \\ \mu \\ \nu \end{bmatrix} = \frac{1}{\sqrt{1+p^2+q^2}} \begin{bmatrix} -p \\ -q \\ 1 \end{bmatrix}$$

we obtain

$$\nu = \frac{1}{\sqrt{1+p^2+q^2}}.$$

Thus, we can write

$$s_1 = \nu h_x (\alpha_1 + p\gamma_1)^2 + \nu h_y (\beta_1 + q\gamma_1)^2. \tag{4.26}$$

Similarly (Fig. 4.26c) we arrive at the expression:

$$s_2 \Delta u = s_{(x)} \Delta u \mathbf{e}_2^* \mathbf{e}_{(x)} \mathbf{e}_2^* \mathbf{e}_{(x)} + s_{(y)} \Delta u \mathbf{e}_2^* \mathbf{e}_{(y)} \mathbf{e}_2^* \mathbf{e}_{(y)}$$

or

$$s_2 = \nu h_x (\alpha_2 + p\gamma_2)^2 + \nu h_y (\beta_2 + q\gamma_2)^2. \tag{4.27}$$

The specific cable forces of the net of principal curvatures satisfy the equation

$$-p_n = s_1 G_1 + s_2 G_2, \tag{4.28}$$

where G_1 and G_2 are the values of the two principal curvatures:

$$G_1 = \frac{1}{R_1} = \nu (r\alpha_1^2 + 2s\alpha_1\beta_1 + t\beta_1^2),$$

$$G_2 = \frac{1}{R_2} = \nu (r\alpha_2^2 + 2s\alpha_2\beta_2 + t\beta_2^2),$$

and p_n is the specific value of the load in the normal direction, referred to the unit area of the net.

If the specific value of the vertical load of the original rectangular cable net related to the x, y plane is p_z, then this load intensity, related to the net surface, has the following component in the normal direction:

$$p_n = \nu^2 p_z.$$

Moreover:

$$-p_z = h_x r + h_y t. \tag{4.29}$$

Example 4.8. The coordinates of the joints of the net used in Example 5.2 of Chapter 5 (see p. 155) appear in Table 5.1. The data of the net is indicated in Fig. 5.3. We now proceed to describe how the data for point 4, 5 of the net, included in Table 5.2, was obtained.

$a = 2.5;\quad b = 2.0;\quad h_x = 10/2 = 5;\quad h_y = 5/2.5 = 2;\quad p = -0.055\,746;$

$q = 0.169\,986;\quad r = 0.007\,346;\quad s = -0.007\,374;\quad t = -0.018\,366.$

Since the net discussed in this example is unloaded, we have

$$-p_z = h_x r + h_y t = 5 \times 0.007\,346 - 2 \times 0.018\,366 = 0.$$

Moreover,

$$\nu = \frac{1}{\sqrt{1+p^2+q^2}} = 0.984\,373$$

$$\mathbf{e}_1 = \begin{bmatrix} \alpha_1 \\ \beta_1 \\ \gamma_1 \end{bmatrix} = \begin{bmatrix} 0.264\,102 \\ 0.953\,180 \\ 0.147\,304 \end{bmatrix}; \quad \mathbf{e}_2 = \begin{bmatrix} 0.962\,933 \\ -0.251\,905 \\ -0.096\,498 \end{bmatrix}$$

$s_1 = 2.206\,202 \qquad s_2 = 4.756\,610$

$G_1 = -0.019\,576 \quad G_2 = 0.009\,079.$

Of course, the normal component of the surface load is also zero, i.e.

$$-p_n = s_1 G_1 + s_2 G_2 = 0.$$

For the same net surface, the rectangular cable net and the cable net of principal curvatures, both of infinitesimal density, will be in a state of equilibrium under different surface loads. If, however, each of the two net types is unloaded, then the net of principal curvatures of infinitesimal density constructed with the aid of the rectangular cable net of infinitesimal density — assuming that the relationship between the cable forces, i.e. between h_x, h_y and s_1, s_2, corresponds to expressions (4.26) and (4.27) —, will be in equilibrium for the case of identical net surfaces. In the case of loaded nets, it is necessary to recall that (4.28) is valid only for loads perpendicular to the surface while (4.29) only holds for vertical loads; furthermore, in the case of identical net surfaces, the relationship

$$p_n = v^2 p_z$$

holds between p_n appearing in (4.28) and p_z occurring in (4.29). It follows from this that under the same load, the two types of net will be in equilibrium in the case of different net surfaces. A method for determining the equilibrium shape of the net belonging to the changed load will be shown in Chapter 5. Here it is sufficient to stress that in the case of nets which are not too steep, the difference between the loads of the two net types belonging to the same net surface is not so great that it influences significantly the design of the net shape. If, for instance, the maximum inclinations on the net surface are of the order of 20% (this is already rather significant), then there may be a maximum 8% difference between p_n and p_z.

What was said in the foregoing is perfectly valid for the cable net of infinitesimal density. The cables of the net of principal curvatures are spatial polygons fitting to the points lying on the lines of principal curvatures, as break points. The cable forces belonging to the polygon sides can be determined only approximately with the aid of specific cable forces s_1, s_2 belonging to the net of infinitesimal density; this is achieved, for instance, by multiplying the cable force intensity belonging to the middle point of the polygon side by the average distance measured from the adjacent, "parallel running" polygon sides.

4.6.3. Construction of the Geodetic Net

The surface curve having the shortest arc-length, connecting two arbitrary points of the surface defined by the rectangular cable net, is a geodetic line passing through the two points. In other words, the line connecting surface points A and B is a geodetic line if the integral

$$I = \int_A^B \sqrt{dx^2+dy^2+dz^2}$$

is a minimum along the line connecting points A and B, running on the surface $z=z(x, y)$. Because of the equality

$$dz = \frac{\partial z}{\partial x} dx + \frac{\partial z}{\partial y} dy = p\, dx + q\, dy,$$

we can write

$$I = \int_{x_A}^{x_B} \sqrt{(1+p^2)+2pqy'+(1+q^2)y'^2}\,\mathrm{d}x; \quad \left(y' = \frac{\mathrm{d}y}{\mathrm{d}x}\right), \tag{4.30}$$

$$I = \int_{y_A}^{y_B} \sqrt{(1+q^2)+2pqx'+(1+p^2)x'^2}\,\mathrm{d}y; \quad \left(x' = \frac{\mathrm{d}x}{\mathrm{d}y}\right). \tag{4.31}$$

The solution of Euler's differential equation of the second order

$$F_y - \frac{\mathrm{d}}{\mathrm{d}x} F_{y'} = 0 \tag{4.32}$$

or

$$F_x^* - \frac{\mathrm{d}}{\mathrm{d}y} F_{x'}^* = 0 \tag{4.33}$$

corresponding to the variational problem

$$I = \int F(x, y, y')\,\mathrm{d}x = \min!$$

or

$$I = \int F^*(y, x, x')\,\mathrm{d}y = \min!$$

gives the equation of the geodetic lines. An infinite number of geodetic lines pass through every point of the surface, but in general only one geodetic line passes through two points. (The geodetic lines of a plane are the straight lines, and only one straight line can be drawn through two points on a plane). If it is assumed that the intersecting cables of the net lie on each other without friction, then the properly pretensioned cables of the net, not fixed to each other, run along the geodetic lines as defined above.

Since Eqs (4.30)–(4.33) are unsuitable for the construction of the geodetic lines of the surface, a very simple numerical procedure — suitable for our purpose but also refinable according to arbitrary requirements of accuracy — has been worked out for the solution to the problem. A computer program for the proposed method can easily be written.

Thus, our problem is to construct a line starting from point A on the ground plan, with a tangent having an angle of inclination φ_A, and defined by the functions $f(x)$ or $f^*(y)$; this line runs into point B which is a distance d from point A in the ground plan, in such a manner that $f(x)$ or $f^*(y)$ should be the ground-plan projection of the surface curve; this, then connects two points of the surface (in the verticals of A and B) and has a minimum length measured on that surface.

The problem is solved by two iterations looped into each other, and the procedure can be refined to arbitrary accuracy. The required geodetic line is constructed in ground plan. Integral I is determined by numerical integration calculated on the surface on the basis of (4.30) or (4.31). The ground plan of the geodetic line is produced from those line sections joining each other whose distance between their starting and end points is equal to length d. The angle of inclination characterizing the tangent of the first line section

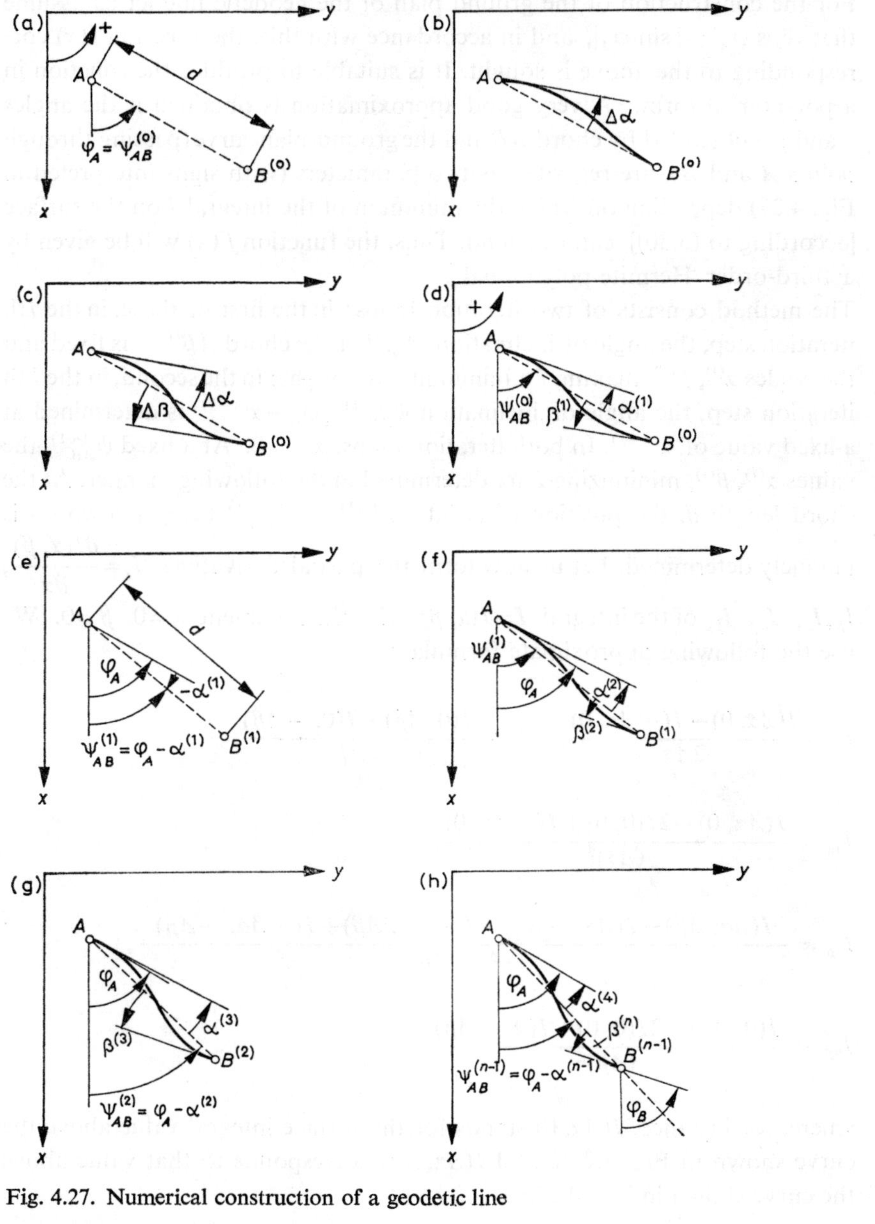

Fig. 4.27. Numerical construction of a geodetic line

at the starting point (point A in Fig. 4.27a) is given, while the tangent of its end point is determined by the method itself. The starting tangent of every further line section coincides with the end tangent of the previous one. The method is the same for every line section; thus, it suffices to discuss it in detail only for the first one.

For the construction of the ground plan of the geodetic line let us assume that $|\cos \varphi_A| \geqslant |\sin \varphi_A|$, and in accordance with this, the function $f(x)$ corresponding to the above is sought. It is suitable to produce the function in a polynomial form. A very good approximation is obtained if the angles α and β, subtended by chord AB and the ground-plan curve (passing through points A and B), are regarded as two parameters (with signs interpreted in Fig. 4.27) depending on which the minimum of the integral I on the surface [according to (5.30)], can be found. Thus, the function $f(x)$ will be given by a third-order Hermite-polynomial.

The method consists of two iteration loops: in the first of these, in the ith iteration step, the angle of inclination $\psi_{AB}^{(i-1)}$ of the chord $AB^{(i-1)}$ is fixed and the angles $\alpha^{(i)}$, $\beta^{(i)}$, making I a minimum are sought; in the second, in the kth iteration step, the angle of inclination $\psi_{AB}^{(k-1)}=\varphi_A-\alpha^{(k-1)}$ is determined at a fixed value of $\alpha^{(k-1)}$. In both iteration loops, $\alpha^{(0)}=0$. At a fixed $\psi_{AB}^{(i-1)}$, the values $\alpha^{(i)}$, $\beta^{(i)}$, minimizing I are determined in the following manner. At the chord length d, the position of point $B^{(i-1)}$ — $\psi_{AB}^{(i-1)}$ being known — is uniquely determined. Let us now form the partial derivatives $I_\alpha=\dfrac{\partial I(\alpha, \beta)}{\partial \alpha}$, I_β, $I_{\alpha\alpha}$, $I_{\alpha\beta}$, $I_{\beta\beta}$ of the integral $I=I(\alpha, \beta)$ with the argument $\alpha=0$, $\beta=0$. We use the following approximate formulae:

$$I_\alpha \approx \frac{I(\Delta\alpha, 0)-I(-\Delta\alpha, 0)}{2\Delta\alpha}; \quad I_\beta \approx \frac{I(0, \Delta\beta)-I(0, -\Delta\beta)}{2\Delta\beta},$$

$$I_{\alpha\alpha} \approx \frac{I(\Delta\alpha, 0)-2I(0, 0)+I(-\Delta\alpha, 0)}{(\Delta\alpha)^2},$$

$$I_{\alpha\beta} \approx \frac{I(\Delta\alpha, \Delta\beta)-I(\Delta\alpha, -\Delta\beta)-I(-\Delta\alpha, \Delta\beta)+I(-\Delta\alpha, -\Delta\beta)}{4\Delta\alpha\, \Delta\beta},$$

$$I_{\beta\beta} \approx \frac{I(0, \Delta\beta)-2I(0, 0)+I(0, -\Delta\beta)}{(\Delta\beta)^2},$$

where, for instance, $I(\Delta\alpha, 0)$ stands for the surface integral value above the curve shown in Fig. 4.27b, and $I(\Delta\alpha, \Delta\beta)$ corresponds to that value above the curve shown in Fig. 4.27c

Having obtained these partial derivatives, the solution of the system of equations

$$\left.\begin{aligned} I_{\alpha\alpha}^{(i)}\alpha_1^{(i)}+I_{\alpha\beta}^{(i)}\beta_1^{(i)} &= -I_{\alpha}^{(i)} \\ I_{\alpha\beta}^{(i)}\alpha_1^{(i)}+I_{\beta\beta}^{(i)}\beta_1^{(i)} &= -I_{\beta}^{(i)} \end{aligned}\right|_{\substack{\alpha=0\\ \beta=0}}$$

gives the first approximation for $\alpha^{(i)}$, $\beta^{(i)}$. The second approximation of $\alpha^{(i)}$, $\beta^{(i)}$ is given by the solution of the equation system

$$\left.\begin{aligned} I_{\alpha\alpha}^{(i)}(\alpha_2^{(i)}-\alpha_1^{(i)})+I_{\alpha\beta}^{(i)}(\beta_2^{(i)}-\beta_1^{(i)}) &= -I_{\alpha}^{(i)} \\ I_{\alpha\beta}^{(i)}(\alpha_2^{(i)}-\alpha_1^{(i)})+I_{\beta\beta}^{(i)}(\beta_2^{(i)}-\beta_1^{(i)}) &= -I_{\beta}^{(i)} \end{aligned}\right|_{\substack{\alpha=\alpha_1^{(i)}\\ \beta=\beta_1^{(i)}}}.$$

The values $\alpha_\nu^{(i)}$, $\beta_\nu^{(i)}$ obtained in the νth approximation are accepted as the actual values of $\alpha^{(i)}$, $\beta^{(i)}$ if, given some prescribed small number ε,

$$|\alpha_\nu^{(i)}-\alpha_{\nu-1}^{(i)}| \leqslant \varepsilon,$$

and

$$|\beta_\nu^{(i)}-\beta_{\nu-1}^{(i)}| \leqslant \varepsilon.$$

As a result of the first cycle of the first iteration loop, the ground plan curve plotted in Fig. 4.27d is obtained. As a second step of the second loop, the chord angle of inclination

$$\psi_{AB}^{(1)} = \varphi_A-\alpha^{(1)}$$

and point $B^{(1)}$ (this being at a distance d from A) are obtained (Fig. 4.27e). Following this, in the second cycle of the first loop the values of $\alpha^{(2)}$ and $\beta^{(2)}$, and thereby the ground plan curves plotted in Figs 4.27f and g, are determined.

After the nth cycle of the first loop the iteration is regarded as completed if

$$|\alpha^{(n)}-\alpha^{(n-1)}| \leqslant \varepsilon.$$

With the angles $\alpha^{(n)}$, $\beta^{(n)}$ obtained at the completion of the iteration (Fig. 4.27h), the angle of inclination characterizing the tangent of the ground plan at point A of the curve constructed on chord $AB^{(n-1)}$ with angle of inclination $\psi_{AB}^{(n-1)}$ will be equal to the prescribed value φ_A, while the surface integral I is a minimum.

The angle of inclination of the starting tangent to the geodetic line section constructed on the next chord with ground plan length d (starting from $B^{(n-1)}$), is given by the expression $\varphi_B=\psi_{AB}^{(n-1)}+\beta^{(n)}$.

The numerical method described above can be refined by shortening the length of the chord d, or by increasing the number of the free parameters of the function giving the arc AB.

The approximate values of the cable forces of the net consisting of cables conducted along the geodetic lines are determined in a manner similar to that used for the net of principle curvatures discussed in Section 4.6.2. The rectangular cable net of infinitesimal density is taken as a basis in this case too. It is assumed that only two geodetic lines are constructed on each surface point. The cable force intensity of the geodetic net, valid in the tangent-plane, can be calculated on the basis of intensities h_x, h_y of the rectangular cable net, in accordance with the principles presented in Section 4.6.2.

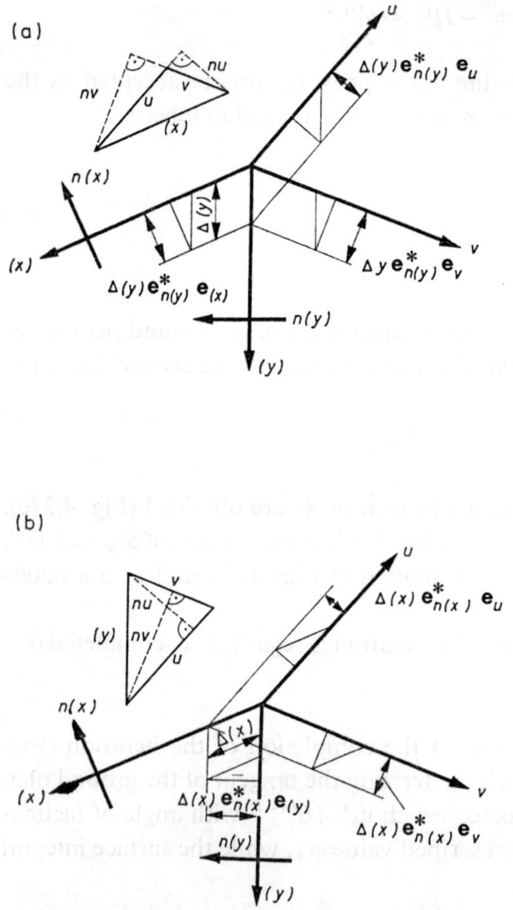

Fig. 4.28. Relations between the cable forces of the rectangular cable net and of the net of geodetic lines.

Let the tangent-plane be specified at some points of the surface, and, as shown in Fig. 4.28, let (x), (y) indicate the cable directions of the rectangular cable net, and u, v the cable directions of the geodetic net. The u, v-directional cable force intensities s_u, s_v corresponding to the (x)-directional cable force intensity $s_{(x)}$ can be constructed according to Fig. 4.28a; those arising under the effect of $s_{(y)}$ can be constructed according to Fig. 4.28b (assuming that u and v are not parallel to x or y):

$$s_u = s_{(x)} \frac{\mathbf{e}^*_{n(y)}\mathbf{e}_{(x)}}{\mathbf{e}^*_{n(y)}\mathbf{e}_u} \frac{\mathbf{e}^*_{nv}\mathbf{e}_{(x)}}{\mathbf{e}^*_{nv}\mathbf{e}_u} + s_{(y)} \frac{\mathbf{e}^*_{n(x)}\mathbf{e}_{(y)}}{\mathbf{e}^*_{n(x)}\mathbf{e}_u} \frac{\mathbf{e}^*_{nv}\mathbf{e}_{(y)}}{\mathbf{e}^*_{nv}\mathbf{e}_u},$$

$$s_v = s_{(x)} \frac{\mathbf{e}^*_{n(y)}\mathbf{e}_{(x)}}{\mathbf{e}^*_{n(y)}\mathbf{e}_v} \frac{\mathbf{e}^*_{nu}\mathbf{e}_{(x)}}{\mathbf{e}^*_{nu}\mathbf{e}_v} + s_{(y)} \frac{\mathbf{e}^*_{n(x)}\mathbf{e}_{(y)}}{\mathbf{e}^*_{n(x)}\mathbf{e}_v} \frac{\mathbf{e}^*_{nu}\mathbf{e}_{(y)}}{\mathbf{e}^*_{nu}\mathbf{e}_v}.$$

By knowing the cable force intensities of the geodetic net, the approximate values of the cable forces of the net can be determined according to the procedure applied in the case of the net of principal curvatures. Once the geometrical data (i.e. joint coordinates) and the cable forces of the net are given, the load carried by the net can be calculated (in accordance with Section 4.2). In general, this load differs from the load of the rectangular cable net which serves as a basis.

5. Exact Calculation of the Change of State of the Net

If the spatial position (x, y, z coordinates) of the joints of the bar net, the tensile (but in the case of a general bar net, tensile or compressive) forces acting in the bars, as well as the position of the edge points of the net are known, then the value of the joint loads compatible with the given geometrical position (and hence the stresses) can be uniquely calculated. It is evident that the variation of the load and the elongations of the bars (the latter may also take place due to a temperature change) are accompanied by the displacement of the joints and by changes in the bar forces. In short: the state of the bar net (i.e. the position of the joints and the values of the bar forces) changes. It is characteristic for the bar net that the change of state is non-linear, because the change of the geometrical position involves the structural change of the equilibrium equation (geometric non-linearity). Besides this, the non-linear character of the change of state can also be caused by the non-linear relationship between the bar elongations and the bar forces (material non-linearity); the latter, however, will not be considered hereinafter. Nevertheless, the proposed calculation method is also capable of taking the physical non-linearity into consideration without any difficulty. First, the investigation of the change of state of the bar net is illustrated on a net stretched on a rigid edge. The purpose of this is simply to outline the relationships in as simple and clear a manner as possible. Bar nets connected to an elastic structure (behaving linearly) are discussed separately. We shall not be concerned with problems of stability during this discussion. For a treatment of the latter, the methods discussed in [25] are recommended.

5.1. Change of State of a Bar Net Fitted to a Rigid Edge

The joint coordinates and the bar forces are called the *state characteristics* of the bar net. The change of state is investigated with the aid of the joint equilibrium equations and the displacement-deformation equations.

5.1.1. Equilibrium Equations

In Section 4.2 the equilibrium equation of the inner joints of the net stretched on a rigid edge was given; this [see Eq. (4.3)] is a matrix equation of the form

$$\mathbf{G}^*\mathbf{s}+\mathbf{p} = 0. \tag{5.1}$$

(For the sake of simplicity, it is assumed that this equation contains only the equilibrium conditions for the inner joints).

Let us examine what happens if the joint loads of the elastic bar net stretched on a rigid edge change by an infinitesimally small load increment. Let us assume that the bar forces and the joint coordinates also undergo an infinitesimally small change due to the change of the load, that is, the equilibrium equation corresponding to the change of load can be written in the form

$$(\mathbf{G}^*+d\mathbf{G}^*)(\mathbf{s}+d\mathbf{s})+\mathbf{p}+d\mathbf{p} = 0. \tag{5.2}$$

Matrix $\mathbf{G}^*$ is exclusively a function of the joint coordinates, which can be defined by a hypervector $\mathbf{r}$:

$$\mathbf{r} = \begin{bmatrix} \mathbf{r}_1 \\ \mathbf{r}_2 \\ \vdots \\ \mathbf{r}_\nu \end{bmatrix}; \quad \mathbf{r}_i = \begin{bmatrix} x_i \\ y_i \\ z_i \end{bmatrix}: \quad (i = 1, 2, \ldots, \nu).$$

The change of the geometrical matrix $\mathbf{G}^* = \mathbf{G}^*(\mathbf{r})$ in the infinitesimally small neighbourhood of the basic state can be approximated by the expression

$$\mathbf{G}^*+d\mathbf{G}^* = \mathbf{G}^*+\frac{\partial \mathbf{G}^*}{\partial \mathbf{r}}\,d\mathbf{r}.$$

Thus, the equilibrium equation for the changed load (5.2) can also be written in the following form:

$$\mathbf{G}^*\mathbf{s}+\mathbf{G}^*d\mathbf{s}+\left(\frac{\partial \mathbf{G}^*}{\partial \mathbf{r}}\,d\mathbf{r}\right)\mathbf{s}+\left(\frac{\partial \mathbf{G}^*}{\partial \mathbf{r}}\,d\mathbf{r}\right)d\mathbf{s}+\mathbf{p}+d\mathbf{p} = 0.$$

If the second-order terms (containing the products of two differentials) are omitted in this equation and the equality (5.1) is also taken into consideration, the equation

$$\mathbf{G}^*\,d\mathbf{s}+\left(\frac{\partial \mathbf{G}^*}{\partial \mathbf{r}}\,d\mathbf{r}\right)\mathbf{s}+d\mathbf{p} = 0$$

describing the change of the state of equilibrium is obtained. Vector $\mathbf{s}$ belonging to the initial state of equilibrium remains unchanged during the change of state; consequently it figures as a constant multiplying factor when the derivatives of $\mathbf{G}^*$ are formed. Thus, we can write that

$$\left(\frac{\partial \mathbf{G}^*}{\partial \mathbf{r}} \mathrm{d}\mathbf{r}\right)\mathbf{s} = \frac{\partial \mathbf{G}^*\mathbf{s}}{\partial \mathbf{r}} \mathrm{d}\mathbf{r} = \mathbf{D}\,\mathrm{d}\mathbf{r}.$$

Finally, the equilibrium equation for the load change is obtained in the following form:

$$\mathbf{D}\,\mathrm{d}\mathbf{r} + \mathbf{G}^*\,\mathrm{d}\mathbf{s} + \mathrm{d}\mathbf{p} = 0. \tag{5.3}$$

The matrix $\mathbf{D}$ contained in this equation — which is a function of the bar forces $\mathbf{s}$ of the initial state and of the change in the geometrical matrix, i.e. $\mathbf{D}=\mathbf{D}(\mathbf{r}, \mathbf{s})$ — is called the *complementary rigidity matrix* belonging to the initial state of the bar net. Matrix $\mathbf{D}$ is a square matrix of order $3\nu \times 3\nu$. It can also be said that $\mathbf{D}$ is a hypermatrix consisting of a number ν^2 of square blocks of order 3×3. The blocks $\mathbf{D}_{j,k}$ of the hypermatrix

$$\mathbf{D} = [\mathbf{D}_{j,k}] \quad (j, k = 1, 2, \ldots, \nu) \tag{5.4}$$

can be produced as the derivatives of the vector elements of the hypervector $\mathbf{G}^*\mathbf{s}$ having the following form:

$$\sum_{\sigma=1}^{\nu_0} \vartheta_{j,\sigma} \mathbf{e}_{j,\sigma} s_{j,\sigma} \quad (j = 1, 2, \ldots, \nu).$$

(In the above expression — as already stated earlier —
ν_0 is the total number of joints in the bar net,
ν is the number of inner joints, and the value of
$\vartheta_{j,\sigma}$ is either 1, or 0, depending on whether joints j and σ are connected by a bar or not.) Block $\mathbf{D}_{j,k}$ is built up of elementary derivatives

$$\frac{\partial}{\partial \mathbf{r}_j} \mathbf{e}_{j,\sigma} = \left[\frac{\partial}{\partial x_j} \mathbf{e}_{j,\sigma};\ \frac{\partial}{\partial y_j} \mathbf{e}_{j,\sigma};\ \frac{\partial}{\partial z_j} \mathbf{e}_{j,\sigma}\right]$$

(see the Appendix for a more detailed treatment) in which

$$\mathbf{e}_{j,\sigma} = \frac{\mathbf{r}_{j,\sigma}}{(\mathbf{r}^*_{j,\sigma}\mathbf{r}_{j,\sigma})^{1/2}} = \frac{\mathbf{r}_{j,\sigma}}{l_{j,\sigma}}; \quad \mathbf{r}_{j,\sigma} = \mathbf{r}_\sigma - \mathbf{r}_j = \begin{bmatrix} x_\sigma - x_j \\ y_\sigma - y_j \\ z_\sigma - z_j \end{bmatrix}.$$

Now,

$$\frac{\partial}{\partial x_j} \mathbf{e}_{j,\sigma} = \frac{\partial}{\partial x_j} \frac{\mathbf{r}_{j,\sigma}}{l_{j,\sigma}} = \frac{\left(\frac{\partial}{\partial x_j} \mathbf{r}_{j,\sigma}\right) l_{j,\sigma} - \mathbf{r}_{j,\sigma}\left(\frac{\partial}{\partial x_j} l_{j,\sigma}\right)}{l^2_{j,\sigma}},$$

and since

$$\frac{\partial}{\partial x_j}\mathbf{r}_{j,\sigma} = \begin{bmatrix} -1 \\ 0 \\ 0 \end{bmatrix} = -\mathbf{e}_x$$

$$\frac{\partial}{\partial x_j} l_{j,\sigma} = \frac{\partial}{\partial x_j}(\mathbf{r}^*_{j,\sigma}\mathbf{r}_{j,\sigma})^{1/2} =$$

$$= \frac{1}{2}(\mathbf{r}^*_{j,\sigma}\mathbf{r}_{j,\sigma})^{-1/2}\left(\left(\frac{\partial}{\partial x_j}\mathbf{r}^*_{j,\sigma}\right)\mathbf{r}_{j,\sigma} + \mathbf{r}^*_{j,\sigma}\left(\frac{\partial}{\partial x_j}\mathbf{r}_{j,\sigma}\right)\right) = -\mathbf{e}^*_{j,\sigma}\mathbf{e}_x,$$

we have

$$\frac{\partial}{\partial x_j}\mathbf{e}_{j,\sigma} = \frac{-\mathbf{e}_x l_{j,\sigma} + \mathbf{r}_{j,\sigma}\mathbf{e}^*_{j,\sigma}\mathbf{e}_x}{l^2_{j,\sigma}} = \frac{1}{l_{j,\sigma}}(\mathbf{e}_{j,\sigma}\mathbf{e}^*_{j,\sigma} - \mathbf{E})\mathbf{e}_x,$$

and, likewise,

$$\frac{\partial}{\partial y_j}\mathbf{e}_{j,\sigma} = \frac{1}{l_{j,\sigma}}(\mathbf{e}_{j,\sigma}\mathbf{e}^*_{j,\sigma} - \mathbf{E})\mathbf{e}_y,$$

$$\frac{\partial}{\partial z_j}\mathbf{e}_{j,\sigma} = \frac{1}{l_{j,\sigma}}(\mathbf{e}_{j,\sigma}\mathbf{e}^*_{j,\sigma} - \mathbf{E})\mathbf{e}_z.$$

Knowing that

$$\mathbf{E} = [\mathbf{e}_x\mathbf{e}_y\mathbf{e}_z] = \begin{bmatrix} 1 & 0 & 0 \\ 0 & 1 & 0 \\ 0 & 0 & 1 \end{bmatrix}$$

we can write

$$\frac{\partial}{\partial \mathbf{r}_j}\mathbf{e}_{j,\sigma} = \frac{1}{l_{j,\sigma}}(\mathbf{e}_{j,\sigma}\mathbf{e}^*_{j,\sigma} - \mathbf{E}).$$

Since $\mathbf{e}_{\sigma,j} = -\mathbf{e}_{j,\sigma}$ we have

$$\frac{\partial}{\partial \mathbf{r}_\sigma}\mathbf{e}_{j,\sigma} = \frac{\partial}{\partial \mathbf{r}_\sigma}\mathbf{e}_{\sigma,j} = -\frac{1}{l_{j,\sigma}}(\mathbf{e}_{j,\sigma}\mathbf{e}^*_{j,\sigma} - \mathbf{E}).$$

Consequently, if $j=k$, block $\mathbf{D}_{j,k}$ is given by

$$\left.\begin{aligned} &\mathbf{D}_{j,j} = \sum_{\sigma=1}^{\nu_0}(\mathbf{e}_{j,\sigma}\mathbf{e}^*_{j,\sigma} - \mathbf{E})\frac{s_{j,\sigma}}{l_{j,\sigma}}\vartheta_{j,\sigma} \\ &\text{while otherwise} \qquad\qquad (j, k = 1, 2, \ldots, \nu) \\ &\mathbf{D}_{j,k} = (\mathbf{E} - \mathbf{e}_{j,k}\mathbf{e}^*_{k,j})\frac{s_{j,k}}{l_{j,k}}\vartheta_{j,k} = \mathbf{D}_{k,j}. \end{aligned}\right\} \tag{5.5}$$

(It is immediately apparent from the expression for blocks $\mathbf{D}_{j,k}$ that $\mathbf{D}=\mathbf{D}^*$). The formation of the complementary rigidity matrix $\mathbf{D}$ is discussed in detail (and is also illustrated numerically) in Example 5.1 of Section 5.1.4.

5.1.2. Displacement Equations

Let us assume that the bars of the bar net behave in a perfectly elastic manner both under tension and compression. Thus, the length of bar j, k, in the case of an infinitesimally small change of the scalar index number $s_{j,k}$ of the bar force, changes by

$$\frac{l_{j,k}}{EA_{j,k}}\,\mathrm{d}s_{j,k} = F_{j,k}\mathrm{d}s_{j,k},$$

where $l_{j,k}$ is the length of the bar j, k belonging to the unloaded (unstressed) state, $A_{j,k}$ is the cross-sectional area of the bar, E is the modulus of elasticity, and

$$F_{j,k} = \frac{l_{j,k}}{EA_{j,k}} \quad \text{is the flexibility of the bar.}$$

Let us also prescribe for bar j, k an elongation, $\mathrm{d}t_{j,k}$, independent of the variation of the bar force (e.g. due to a temperature change). Let us assume that the change of the state of equilibrium is accompanied by the change of the position vectors $\mathrm{d}\mathbf{r}_j$, $\mathrm{d}\mathbf{r}_k$ belonging to the end points of the bar j, k.
Due to the infinitesimally small change of position of the end points of the bar j, k, the length of the bar undergoes a change given by the scalar product

$$\mathbf{e}^*_{j,k}(\mathrm{d}\mathbf{r}_k-\mathrm{d}\mathbf{r}_j) = \mathbf{e}^*_{j,k}\mathrm{d}\mathbf{r}_{j,k}.$$

This change of length must be equal to the sum of the elastic elongation and the elongation due to other effects. Thus, the equation expressing the compatibility of the displacement of the bar ends and the elongation of the bar can be written in the form

$$\mathbf{e}^*_{j,k}(\mathrm{d}\mathbf{r}_k-\mathrm{d}\mathbf{r}_j) = F_{j,k}\,\mathrm{d}s_{j,k}+\mathrm{d}t_{j,k},$$

or

$$\mathbf{e}^*_{j,k}\,\mathrm{d}\mathbf{r}_j+\mathbf{e}^*_{k,j}\,\mathrm{d}\mathbf{r}_k+F_{j,k}\,\mathrm{d}s_{j,k}+\mathrm{d}t_{j,k} = 0,$$

or, in a more condensed form:

$$[\mathbf{e}^*_{j,k}\mathbf{e}^*_{k,j}]\begin{bmatrix}\mathrm{d}\mathbf{r}_j\\ \mathrm{d}\mathbf{r}_k\end{bmatrix}+F_{j,k}\,\mathrm{d}s_{j,k}+\mathrm{d}t_{j,k} = 0.$$

Such an equation holds for every bar of the net. The complete set of equations constitutes the displacement-deformation compatibility matrix equation of the bar net:

$$\mathbf{G}\,\mathbf{dr}+\mathbf{F}\,\mathbf{ds}+\mathbf{dt}=0. \tag{5.6}$$

In the case shown in Fig. 5.1 and assuming the edge points to be fixed

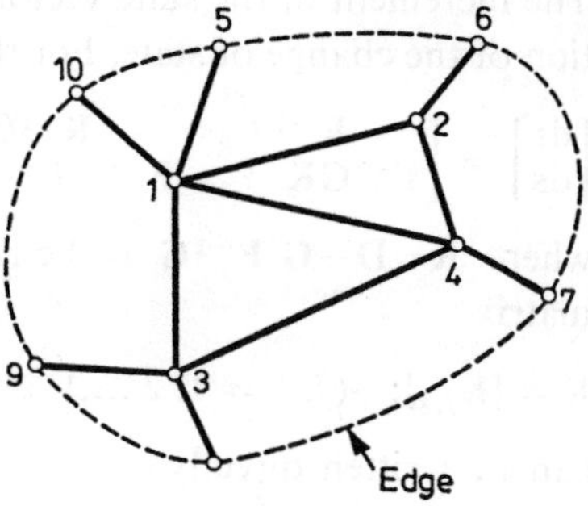

5.1. Single-layer bar net of general shape

($\mathbf{dr}_j=0$, if $j=5, 6, \ldots, 10$), the elements of the matrices **G** and **F** in the equation are:

$$\mathbf{G}=\begin{bmatrix} \mathbf{e}^*_{1,2} & \mathbf{e}^*_{2,1} & & \\ \mathbf{e}^*_{1,3} & & \mathbf{e}^*_{3,1} & \\ \mathbf{e}^*_{1,4} & & & \mathbf{e}^*_{4,1} \\ \mathbf{e}^*_{1,5} & & & \\ \mathbf{e}^*_{1,10} & & & \\ & \mathbf{e}^*_{2,4} & & \mathbf{e}^*_{4,2} \\ & \mathbf{e}^*_{2,6} & & \\ & & \mathbf{e}^*_{3,4} & \mathbf{e}^*_{4,3} \\ & & \mathbf{e}^*_{3,8} & \\ & & \mathbf{e}^*_{3,9} & \\ & & & \mathbf{e}^*_{4,7} \end{bmatrix}$$

$$\mathbf{F}=\langle F_{1,2}F_{1,3}F_{1,4}F_{1,5}F_{1,10}F_{2,4}F_{2,6}F_{3,4}F_{3,8}F_{3,9}F_{4,7}\rangle.$$

5.1.3. Equation for the Change of State

The equilibrium equation (5.3) and the displacement equation (5.6) of the bar net can be summarized in a single hypermatrix equation

$$\begin{bmatrix}\mathbf{D} & \mathbf{G}^* \\ \mathbf{G} & \mathbf{F}\end{bmatrix}\begin{bmatrix}\mathbf{dr} \\ \mathbf{ds}\end{bmatrix}+\begin{bmatrix}\mathbf{dp} \\ \mathbf{dt}\end{bmatrix}=0. \tag{5.7}$$

Equation (5.7) is called the *equation of the change of state of the bar net.* Hereinafter, vector $\begin{bmatrix}\mathbf{r}\\ \mathbf{s}\end{bmatrix}$ will be called the state vector and vector $\begin{bmatrix}\mathrm{d}\mathbf{r}\\ \mathrm{d}\mathbf{s}\end{bmatrix}$ will be called the vector of the infinitesimally small change of state.
$\begin{bmatrix}\mathbf{p}\\ \mathbf{t}\end{bmatrix}$ and $\begin{bmatrix}\mathrm{d}\mathbf{p}\\ \mathrm{d}\mathbf{t}\end{bmatrix}$ are the vectors denoting the load and the change of load, respectively.
The increment of the state vector can be expressed explicitly from the equation of the change of state, For the case $|\mathbf{F}|\neq 0$

$$\begin{bmatrix}\mathrm{d}\mathbf{r}\\ \mathrm{d}\mathbf{s}\end{bmatrix}=\begin{bmatrix}-\mathbf{K}^{-1} & \mathbf{K}^{-1}\mathbf{G}^*\mathbf{F}^{-1}\\ \mathbf{F}^{-1}\mathbf{G}\mathbf{K}^{-1} & -\mathbf{F}^{-1}-\mathbf{F}^{-1}\mathbf{G}\mathbf{K}^{-1}\mathbf{G}^*\mathbf{F}^{-1}\end{bmatrix}\begin{bmatrix}\mathrm{d}\mathbf{p}\\ \mathrm{d}\mathbf{t}\end{bmatrix}, \tag{5.8}$$

where $\mathbf{K}=\mathbf{D}-\mathbf{G}^*\mathbf{F}^{-1}\mathbf{G}$ is the rigidity matrix of the bar net. The blocks of matrix

$\mathbf{K}=[\mathbf{K}_{j,k}];\quad (j,\, k=1, 2\,\ldots,)$

can be written directly.
If $k=j$,

$$\left.\begin{aligned}\mathbf{K}_{j,j}&=-\sum_{\sigma=1}^{v_0}\left(\mathbf{E}+\left(\frac{EA_{j,\sigma}}{s_{j,\sigma}}-1\right)\mathbf{e}_{j,\sigma}\mathbf{e}^*_{j,\sigma}\right)\frac{s_{j,\sigma}}{l_{j,\sigma}}\vartheta_{j,\sigma}\\ &\text{and otherwise}\\ \mathbf{K}_{j,k}&=\left(\mathbf{E}-\left(\frac{EA_{j,k}}{s_{j,k}}+1\right)\mathbf{e}_{j,k}\mathbf{e}^*_{j,k}\right)\frac{s_{j,k}}{l_{j,k}}\vartheta_{j,k}\,.\end{aligned}\right\} \tag{5.9}$$

If $\mathbf{F}=0$ but $|\mathbf{D}|\neq 0$, then

$$\begin{bmatrix}\mathrm{d}\mathbf{r}\\ \mathrm{d}\mathbf{s}\end{bmatrix}=\begin{bmatrix}\mathbf{D}^{-1}\mathbf{G}^*(\mathbf{G}\mathbf{D}^{-1}\mathbf{G}^*)^{-1}\mathbf{G}\mathbf{D}^{-1}-\mathbf{D}^{-1} & -\mathbf{D}^{-1}\mathbf{G}^*(\mathbf{G}\mathbf{D}^{-1}\mathbf{G}^*)^{-1}\\ -(\mathbf{G}\mathbf{D}^{-1}\mathbf{G}^*)^{-1}\mathbf{G}\mathbf{D}^{-1} & (\mathbf{G}\mathbf{D}^{-1}\mathbf{G}^*)^{-1}\end{bmatrix}\cdot\begin{bmatrix}\mathrm{d}\mathbf{p}\\ \mathrm{d}\mathbf{t}\end{bmatrix}. \tag{5.8a}$$

5.1.4. Numerical Method of Analysis for the Finite Change of State

The differential equation (5.7) for the change of state of the cable net attached to the rigid edge serves for the exact description of the infinitesimally small change taking place in the neighbourhood of the given state. The equation is apparently linear; in reality, however, it is not, because blocks **D** and **G** of the coefficient matrix depend on the variables **r**, **s**, i.e.

$\mathbf{D}=\mathbf{D}(\mathbf{r},\mathbf{s});\quad \mathbf{G}=\mathbf{G}(\mathbf{r}).$

We do not endeavour to obtain an analytical solution for the problem of finding the initial value of the change of state. Instead, a numerical method is presented, which can be regarded as exact because after every step of the approximation it is possible to determine exactly the error in the load, compatible with the state, related to the prescribed load; since this error can be reduced below an arbitrarily prescribed limit, the requirements of the calculation are met.

The requirement concerning accuracy also leads to the outline of the numerical method. Accordingly, the two principal parts of the method are:

1. An approximately exact determination of the change of state due to the load increment.
2. Having the data of the changed state, the *exact* determination of the compatible load and, with knowledge of this, that of the error vector.

An *approximately* exact result is obtained by regarding Eq. (5.7) — which is exact for infinitesimally small changes — as valid for small, but finite, changes too. The characteristics of the initial state, i.e. vectors $\mathbf{r}_{(0)}$, $\mathbf{s}_{(0)}$, matrices $\mathbf{G}_{(0)}$, $\mathbf{D}_{(0)}$, and the load compatible with the state (i.e. vectors $\mathbf{p}_{(0)}, \mathbf{t}_{(0)}$), are regarded as given. It is especially important to pay attention to the calculation of the latter because their computation is the basis of the exact part of the method.

The vector of the joint loads compatible with the given state is determined by the expression

$$\mathbf{p}_{(0)} = -\mathbf{G}^*_{(0)}\mathbf{s}_{(0)}. \tag{5.10}$$

The vector $\mathbf{t}$ of the kinematic loads compatible with the given state is determined by the expression

$$\mathbf{t}_{(0)} = \mathbf{l}_{(0)} - \mathbf{l} - \mathbf{F}\mathbf{s}_{(0)} \tag{5.11}$$

where $\mathbf{l}_{(0)}$ is the vector formed from the bar lengths calculated with the aid of $\mathbf{r}_{(0)}$ and $\mathbf{l}$ is the vector formed from the unstressed bar lengths corresponding to the erection temperature. The flexibility matrix $\mathbf{F}$ is regarded as constant. By knowing the load compatible with the initial state ($\mathbf{p}_{(0)}, \mathbf{t}_{(0)}$), we aim at the calculation of a state $\mathbf{r}, \mathbf{s}$ belonging to a prescribed load $\mathbf{p}, \mathbf{t}$. Let us assume that the load increment

$$\Delta\mathbf{p} = \mathbf{p} - \mathbf{p}_{(0)},$$

$$\Delta\mathbf{t} = \mathbf{t} - \mathbf{t}_{(0)}$$

is small, but finite. This produces a change of state

$$\Delta\mathbf{r} = \mathbf{r} - \mathbf{r}_{(0)},$$

$$\Delta\mathbf{s} = \mathbf{s} - \mathbf{s}_{(0)}$$

that can be determined, on the basis of Eq. (5.7), only with approximate "exactness":

$$\begin{bmatrix} \mathbf{D}_{(0)} & \mathbf{G}^*_{(0)} \\ \mathbf{G}_{(0)} & \mathbf{F} \end{bmatrix} \begin{bmatrix} \Delta\mathbf{r}_{(1)} \\ \Delta\mathbf{s}_{(1)} \end{bmatrix} + \begin{bmatrix} \Delta\mathbf{p} \\ \Delta\mathbf{t} \end{bmatrix} = 0, \tag{5.12}$$

where $\Delta\mathbf{r}_{(1)}$ and $\Delta\mathbf{s}_{(1)}$ stand for the first approximations of $\Delta\mathbf{r}$ and $\Delta\mathbf{s}$. Here, it is suitable to proceed as follows:

1. Let us determine the reduced load

$$\Delta\hat{\mathbf{p}} = \Delta\mathbf{p} - \mathbf{G}_{(0)}\mathbf{F}^{-1}\Delta\mathbf{t}.$$

2. With the aid of equation

$$\mathbf{K}_{(0)}\Delta\mathbf{r}_{(1)} + \Delta\hat{\mathbf{p}} = 0$$

let us calculate the vector $\Delta\mathbf{r}_{(1)}$;

3. Let us determine the value of the reduced kinematic load vector

$$\Delta\hat{\mathbf{t}}_{(1)} = \mathbf{G}^*_{(0)}\Delta\mathbf{r}_{(1)} + \Delta\mathbf{t},$$

and finally

4. the bar force increment:

$$\Delta\mathbf{s}_{(1)} = -\mathbf{F}^{-1}\Delta\hat{\mathbf{t}}_{(1)}.$$

The characteristics of the new state are

$$\mathbf{r}_{(1)} = \mathbf{r}_{(0)} + \Delta\mathbf{r}_{(1)},$$

$$\mathbf{s}_{(1)} = \mathbf{s}_{(0)} + \Delta\mathbf{s}_{(1)},$$

and the coefficients of the equation of the change of state are: $\mathbf{G}_{(1)}$, $\mathbf{D}_{(1)}$. The load vector compatible with the new state is

$$\mathbf{p}_{(1)} = -\mathbf{G}^*_{(1)}\mathbf{s}_{(1)},$$

$$\mathbf{t}_{(1)} = \mathbf{l}_{(1)} - \mathbf{F}\mathbf{s}_{(1)} - \mathbf{l}.$$

The load compatible with this state deviates from the prescribed load $\mathbf{p}$, $\mathbf{t}$ by the error

$$\Delta\mathbf{p}_1 = \mathbf{p} - \mathbf{p}_{(1)},$$

$$\Delta\mathbf{t}_1 = \mathbf{t} - \mathbf{t}_{(1)}.$$

This vector

$$\begin{bmatrix} \Delta\mathbf{p}_{(1)} \\ \Delta\mathbf{t}_{(1)} \end{bmatrix}$$

is called the *error vector* of the new state.

With the knowledge of the error vector the next step of the calculation can proceed. If

(a) $\|\Delta \mathbf{p}_{(1)}\| < \varepsilon_p \|\tilde{\mathbf{p}}\|$,

and $\|\Delta \mathbf{t}_{(1)}\| < \varepsilon_t \|\tilde{\mathbf{t}}\|$

then the calculation is regarded as completed with "relative exactness" $\varepsilon_p, \varepsilon_t$;
(here $\tilde{\mathbf{p}}$ and $\tilde{\mathbf{t}}$ are equal to the actual load or to some other, possible, average load on the structure, such that $\tilde{\mathbf{p}} \neq 0$, $\tilde{\mathbf{t}} \neq 0$).
(b) The prescription under (a) is not fulfilled partly or entirely, but

$\|\Delta \mathbf{p}_{(1)}\| \ll \|\Delta \tilde{\mathbf{p}}\|$,

and

$\|\Delta \mathbf{t}_{(1)}\| \ll \|\Delta \tilde{\mathbf{t}}\|$

(here $\Delta \tilde{\mathbf{p}}$ and $\Delta \tilde{\mathbf{t}}$ are equal either to the actual load increment, or to a fraction of the possible average load on the structure, but always in such a manner that $\Delta \tilde{\mathbf{p}} \neq 0$ and $\Delta \tilde{\mathbf{t}} \neq 0$). Then, regarding the new state as the initial state, the first step is iterated until the error vector

$$\begin{bmatrix} \Delta \mathbf{p}_{(i)} \\ \Delta \mathbf{t}_{(i)} \end{bmatrix}$$

associated with the ith step completely satisfies the conditions under (a);
(c) if neither the condition under (a), nor the one under (b) is fulfilled, then two cases are possible, namely:

(ca) in the case of a reduced load step ($\lambda \Delta \mathbf{p}$, $\lambda \Delta \mathbf{t}$) formed by a suitably chosen multiplicator $\lambda (0 < \lambda < 1)$, the state corresponding to the reduced load step is determined with an exactness prescribed according to condition (a) or (b); then, following this, the missing load part is applied (possibly in several steps);
(cb) even by letting the multiplicator $\lambda \to 0$ it is not possible to determine the new state belonging to the reduced load increment by convergent iteration.

In the case of (cb), the critical point of the stable change of state has been reached. Any further investigation of the bar net should be done by methods suitable for describing the post-critical behaviour — these being beyond the scope of our analyses.
The essential features of the outlined procedure are illustrated in the following simple example.

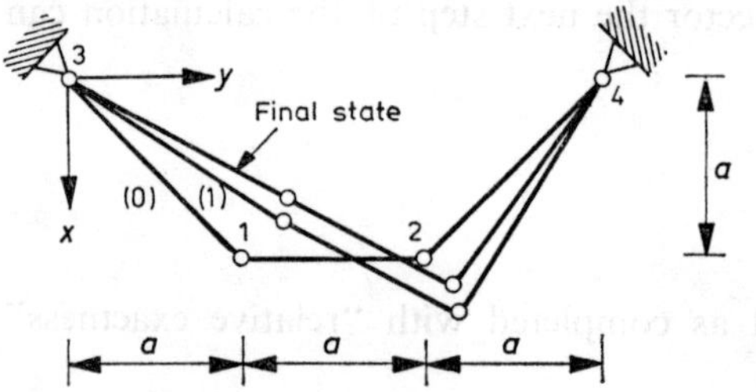

Fig. 5.2. Change of state of a plane bar chain

Example 5.1: The state vectors belonging to the initial state of the plane bar chain depicted in Fig. 5.2. are:

$$\mathbf{r}_{(0)} = \begin{bmatrix} \mathbf{r}_1 \\ \mathbf{r}_2 \end{bmatrix} = a \begin{bmatrix} 1 \\ 1 \\ 1 \\ 2 \end{bmatrix}; \quad \mathbf{s}_{(0)} = H \begin{bmatrix} 1 \\ \sqrt{2} \\ \sqrt{2} \end{bmatrix} = \begin{bmatrix} s_{1,2} \\ s_{1,3} \\ s_{2,4} \end{bmatrix}.$$

Vector $\mathbf{r}_{(0)}$ corresponds to the position marked by (0) as shown in the figure; vector $\mathbf{s}_{(0)}$ was arbitrarily chosen, with the restriction that the *y*-directional components of all three bar forces should be equal to H.
The coefficient matrices corresponding to state (0) are:

$$\mathbf{G}_{(0)} = \begin{bmatrix} \mathbf{e}^*_{1,2} & \mathbf{e}^*_{2,1} \\ \mathbf{e}^*_{1,3} & \\ & \mathbf{e}^*_{2,4} \end{bmatrix} = \begin{bmatrix} 0 & 1 & 0 & -1 \\ -1/\sqrt{2} & -1/\sqrt{2} & & \\ & & -1/\sqrt{2} & 1/\sqrt{2} \end{bmatrix};$$

$\mathbf{F} = 0$ (rigid bars)

$$\mathbf{D}_{(0)} = \begin{bmatrix} \mathbf{D}_{(0)1,1} & \mathbf{D}_{(0)1,2} \\ \mathbf{D}_{(0)2,1} & \mathbf{D}_{(0)2,2} \end{bmatrix}; \quad \mathbf{l}_{(0)} = \mathbf{l} = a \begin{bmatrix} 1 \\ \sqrt{2} \\ \sqrt{2} \end{bmatrix}.$$

$$\mathbf{D}_{(0)1,1} = (\mathbf{e}_{1,2}\mathbf{e}^*_{1,2} - \mathbf{E}) \frac{s_{1,2}}{l_{1,2}} + (\mathbf{e}_{1,3}\mathbf{e}^*_{1,3} - \mathbf{E}) \frac{s_{1,3}}{l_{1,3}} = \begin{bmatrix} -1.5 & 0.5 \\ 0.5 & -0.5 \end{bmatrix} \frac{H}{a}$$

$$\mathbf{D}_{(0)1,2} = (\mathbf{E} - \mathbf{e}_{1,2}\mathbf{e}^*_{1,2}) \frac{s_{1,2}}{l_{1,2}} = \begin{bmatrix} 1 & 0 \\ 0 & 0 \end{bmatrix} \frac{H}{a} = \mathbf{D}_{(0)2,1}$$

$$\mathbf{D}_{(0)2,2} = (\mathbf{e}_{2,1}\mathbf{e}^*_{2,1} - \mathbf{E}) \frac{s_{2,1}}{l_{2,1}} + (\mathbf{e}_{2,4}\mathbf{e}^*_{2,4} - \mathbf{E}) \frac{s_{2,4}}{l_{2,4}} = \begin{bmatrix} -1.5 & -0.5 \\ -0.5 & -0.5 \end{bmatrix} \frac{H}{a}.$$

Thus:

$$\mathbf{D}_{(0)} = \begin{bmatrix} -1.5 & 0.5 & 1 & 0 \\ 0.5 & -0.5 & 0 & 0 \\ 1 & 0 & -1.5 & -0.5 \\ 0 & 0 & -0.5 & -0.5 \end{bmatrix} \frac{H}{a}.$$

The compatible load is

$$\mathbf{p}_{(0)} = -\mathbf{G}^*_{(0)}\mathbf{s}_{(0)} = H\begin{bmatrix}1\\0\\1\\0\end{bmatrix}$$

$\mathbf{t}_{(0)} = \mathbf{l}_{(0)} - \mathbf{F}\mathbf{s}_{(0)} - \mathbf{l} = 0$, (because $\mathbf{l}_{(0)} = \mathbf{l}$ and $\mathbf{F} = 0$).

Let the prescribed load be

$$\mathbf{p} = H\begin{bmatrix}0\\0\\2\\0\end{bmatrix}; \quad \mathbf{t} = 0,$$

thus the load increment is

$$\Delta\mathbf{p} = \mathbf{p} - \mathbf{p}_{(0)} = H\begin{bmatrix}-1\\0\\1\\0\end{bmatrix}; \quad \Delta\mathbf{t} = 0.$$

Equation

$$\begin{bmatrix}\mathbf{D}_{(0)} & \mathbf{G}^*_{(0)}\\ \mathbf{G}_{(0)} & 0\end{bmatrix}\begin{bmatrix}\Delta\mathbf{r}_{(1)}\\ \Delta\mathbf{s}_{(1)}\end{bmatrix} + \begin{bmatrix}\Delta\mathbf{p}\\ \Delta\mathbf{t}\end{bmatrix} = 0$$

now has to be solved according to (5.8a) in the form

$$\Delta\mathbf{s}_{(1)} = \mathbf{H}^{-1}_{(0)}(\Delta\mathbf{t} - \mathbf{G}_{(0)}\mathbf{D}^{-1}_{(0)}\Delta\mathbf{p}); \quad \mathbf{H}_{(0)} = \mathbf{G}_{(0)}\mathbf{D}^{-1}_{(0)}\mathbf{G}^*_{(0)}$$

$$\Delta\mathbf{r}_{(1)} = -\mathbf{D}^{-1}_{(0)}(\mathbf{G}^*_{(0)}\Delta\mathbf{s}_{(1)} + \Delta\mathbf{p})$$

(because $\mathbf{F}=0$):

$$\mathbf{s}_1 = \frac{H\sqrt{2}}{4}\begin{bmatrix}0\\-1\\1\end{bmatrix}; \quad \Delta\mathbf{r}_1 = \frac{a}{4}\begin{bmatrix}-1\\1\\1\\1\end{bmatrix}.$$

The new state vectors are

$$\mathbf{s}_{(1)} = \mathbf{s}_{(0)} + \Delta\mathbf{s}_{(1)} = H\begin{bmatrix}1.0\\1.060\,660\\1.767\,767\end{bmatrix},$$

$$\mathbf{r}_{(1)} = \mathbf{r}_{(0)} + \Delta\mathbf{r}_{(1)} = a\begin{bmatrix}0.75\\1.25\\1.25\\2.25\end{bmatrix}, \quad \mathbf{l}_{(1)} = a\begin{bmatrix}1.118\,034\\1.457\,738\\1.457\,738\end{bmatrix},$$

and the geometrical matrix is:

$$\mathbf{G}_{(1)} = \begin{bmatrix} 0.447\,214 & 0.894\,427 & -0.447\,214 & -0.894\,427 \\ -0.514\,496 & -0.857\,493 & & \\ & & -0.857\,493 & 0.514\,496 \end{bmatrix}.$$

The load vectors compatible with the state are:

$$\mathbf{p}_{(1)} = -\mathbf{G}^*_{(1)}\,\mathbf{s}_{(1)} = H \begin{bmatrix} 0.098\,491 \\ 0.015\,082 \\ 1.963\,062 \\ -0.015\,082 \end{bmatrix},$$

$$\mathbf{t}_{(1)} = \mathbf{l}_{(1)} - \mathbf{l} = a \begin{bmatrix} 0.118\,034 \\ 0.043\,524 \\ 0.043\,524 \end{bmatrix}.$$

The error vectors are:

$$\Delta\mathbf{p}_{(1)} = H \begin{bmatrix} -0.098\,491 \\ -0.015\,082 \\ 0.036\,938 \\ 0.015\,082 \end{bmatrix}, \quad \Delta\mathbf{t}_{(1)} = \mathbf{t} - \mathbf{t}_{(1)} = a \begin{bmatrix} -0.118\,034 \\ -0.043\,524 \\ -0.043\,524 \end{bmatrix}.$$

Continuing the iteration, the end state is reached after four steps, in which

$$\mathbf{s}_{(4)} = \mathbf{s} = H \begin{bmatrix} 1.237\,299 \\ 1.237\,299 \\ 1.797\,896 \end{bmatrix},$$

$$\mathbf{r}_{(4)} = \mathbf{r} = a \begin{bmatrix} 0.656\,781 \\ 1.252\,453 \\ 1.121\,195 \\ 2.138\,071 \end{bmatrix},$$

$$\mathbf{l}_{(4)} = \mathbf{l} = a \begin{bmatrix} 1.0 \\ 1.414\,214 \\ 1.414\,214 \end{bmatrix}.$$

It appears from the figure (and also from the data of the results) that the end state was already adequately approximated after the first iteration step; the latter can also be easily determined by elementary means.
In the next example the change of state of a net constructed according to the lines of principal curvatures is investigated:

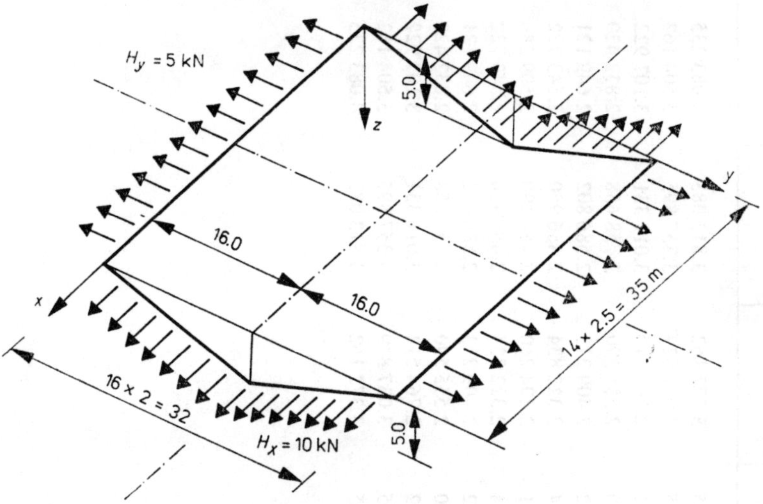

Fig. 5.3. Rectangular net stretched on a rigid edge

Example 5.2. The joints of the net stretched on the rigid edge shown in Fig. 5.3 are unloaded. The net consists of 15 x-directional and 13 y-directional cables. The $z_{j,k}$ coordinates of the rectangular net are summarized in Table 5.1. (Because of the symmetry only half of the data is shown).
The tangents of the principal curvatures belonging to the joints of the cable net rectangular in ground plan, the principal curvatures and the specific cable forces of the cable net of principal curvatures were determined according to the procedure discussed in Sections 4.6.1 and 4.6.2; they are given in Table 5.2. The direction field of the principal curvatures and the construction of the joints of the net of principal curvatures (for a quarter of the entire net) are shown in Fig. 5.4.
We have deliberately constructed the net of principal curvatures by means of few cables in order to demonstrate that the net constructed in this way also fits quite well the surface defined by the original cable net having a rectangular ground plan. When constructing the joints of the cable net of principal curvatures, the lines of principal curvatures $X_1, \ldots, X_5$ and $Y_1, \ldots, Y_5$ (Fig. 5.5) were generated manually, by graphic integration; the x_i, y_i coordinates of the joints $i = 1, 2, \ldots, 49$ of the net of principal curvatures were also determined graphically, i.e. they were measured from the figure. The z_i coordinates were interpolated with the aid of the $z_{j,k}$ coordinates of the rect-

Table 5.1 $z_{j,k}$

k \ j	1	2	3	4	5	6	7	8
1	0.588 054	1.172 830	1.750 540	2.316 121	2.861 635	3.372 162	3.813 083	4.083 235
2	0.553 158	1.100 075	1.633 661	2.144 784	2.620 135	3.037 829	3.357 895	3.504 162
3	0.522 161	1.035 052	1.530 822	1.995 805	2.414 672	2.764 514	3.011 331	3.107 922
4	0.496 584	0.982 679	1.446 850	1.875 647	2 252 350	2.555 590	2.758 658	2.832 420
5	0.477 561	0.943 410	1.384 986	1.788 048	2.135 942	2.409 273	2.586 802	2.649 121
6	0.465 859	0.919 311	1.347 193	1.734 931	2.066 134	2.322 834	2.486 950	2.543 722
7	0.461 912	0.911 192	1.334 492	1.717 146	2.042 891	2.294 260	2.454 190	2.509 286
8	0.465 859	0.919 311	1.347 193	1.734 931	2.066 134	2.322 834	2.486 950	2.543 722
9	0.477 561	0.943 410	1.384 986	1.788 048	2.135 942	2.409 273	2.586 802	2.649 121
10	0.496 584	0.982 679	1.446 850	1.875 647	2.252 350	2.555 590	2.758 658	2.832 420
11	0.522 161	1.035 652	1.530 822	1.995 805	2.414 672	2.764 514	3.011 331	3.107 922
12	0.553 158	1.100 075	1.633 661	2.144 784	2.620 135	3.037 829	3.357 895	3.504 162
13	0.588 054	1.172 830	1.750 540	2.316 121	2.861 635	3.372 162	3.813 083	4.083 235

angular cable net (to be found in Table 5.1) by the expression

$$z_i = z_{j,k}\left(1-\frac{\xi_i}{a}-\frac{\eta_i}{b}+\frac{\xi_i\eta_i}{ab}\right)+z_{j+1,k}\left(\frac{\xi_i}{a}-\frac{\xi_i\eta_i}{ab}\right)+$$

$$+z_{j,k+1}\left(\frac{\eta_i}{b}-\frac{\xi_i\eta_i}{ab}\right)+z_{j+1,k+1}\frac{\xi_i\eta_i}{ab},$$

where $\xi_i=x_i-ja$, $\eta_i=y_i-kb$, and j, k designate that joint of the rectangular cable net which lies nearest to joint i and for which $ja\leqslant x_i$ and $kb\leqslant y_i$. For example, at joint $i=9$ of Fig. 5.4, $j=2$, $k=3$, and because $a=2.5$, $b=2.0$ (see Fig. 5.3), $x_9=5.88$, $y_9=6.55$

$$z_9 = 1.633\,661\left(1-\frac{0.88}{2.5}-\frac{0.55}{2}+\frac{0.88}{2.5}\,\frac{0.55}{2}\right)+$$

$$+1.530\,822\left(\frac{0.88}{2.5}-\frac{0.88}{2.5}\,\frac{0.55}{2}\right)+$$

$$+2.144\,784\left(\frac{0.55}{2}-\frac{0.88}{2.5}\,\frac{0.55}{2}\right)+$$

$$+1.995\,805\,\frac{0.88}{2.5}\,\frac{0.55}{2} = 1.733\,554.$$

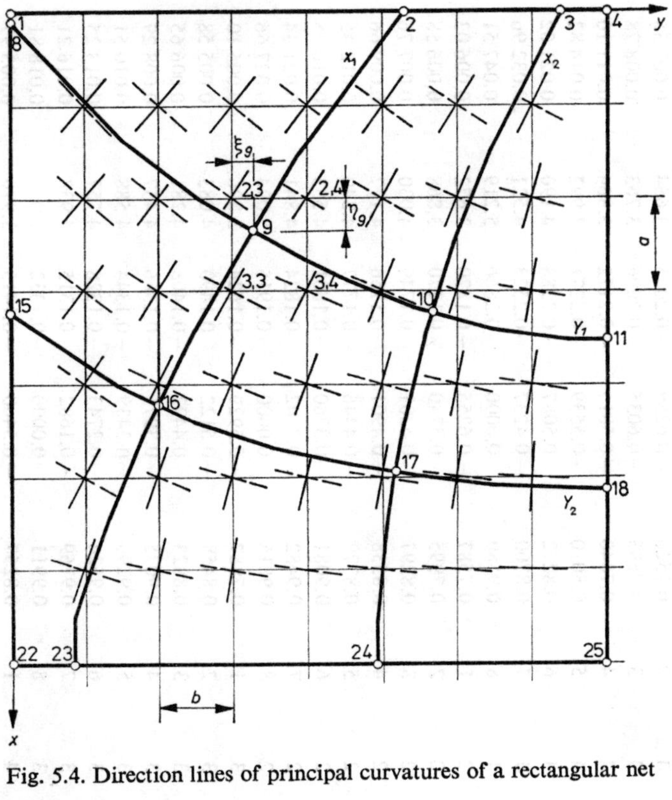

Fig. 5.4. Direction lines of principal curvatures of a rectangular net

Table 5.2. Principal curvatures (G), tangents to principal curvatures ($\alpha\ \beta\ \gamma$), and specific tensile forces (s) of a cable net with rectangular ground plan

j	k	α_1	β_1	γ_1	s_1	G_1	α_2	β_2	γ_2	s_2	G_2
1	1	0.7317	−0.6512	−0.2014	3.557	0.006 69	0.6815	0.7049	0.1969	3.326	−0.007 15
1	2	0.7526	−0.6259	−0.2045	3.664	0.007 38	0.6579	0.7283	0.1919	3.221	−0.008 40
1	3	0.7693	−0.6035	−0.2096	3.763	0.008 78	0.6372	0.7486	0.1832	3.128	−0.010 56
1	4	0.7840	−0.5815	−0.2172	3.864	0.011 10	0.6169	0.7685	0.1696	3.038	−0.014 12
1	5	0.8010	−0.5539	−0.2271	3.992	0.014 82	8.5907	0.7929	0.1497	2.928	−0.020 20
1	6	0.8272	−0.5087	−0.2388	4.190	0.021 02	0.5449	0.8299	0.1198	2.763	−0.031 88
1	7	0.8700	−0.4237	−0.2523	4.521	0.032 96	0.4523	0.8894	0.0661	2.498	−0.059 67
1	8	0.9580	0.0000	−0.2866	5.219	0.047 51	0.0000	1.0000	0.0000	1.916	−0.129 41
2	1	0.7587	−0.6255	−0.1820	3.672	0.006 03	0.6513	0.7338	0.1932	3.223	−0.006 88
2	2	0.7995	−0.5740	−0.1770	3.868	0.006 58	0.6001	0.7763	0.1932	3.032	−0.008 39
2	3	0.8297	−0.5301	−0.1749	4.030	0.007 76	0.5566	0.8094	0.1869	2.878	−0.010 86
2	4	0.8538	−0.4901	−0.1756	4.177	0.009 66	0.5169	0.8382	0.1736	2.744	−0.014 70
2	5	0.8779	−0.4446	−0.1777	4.340	0.012 45	0.4709	0.8690	0.1519	2.602	−0.020 77
2	6	0.9081	−0.3780	−0.1801	4.553	0.016 35	0.4015	0.9081	0.1187	2.421	−0.030 75
2	7	0.9482	−0.2602	−0.1824	4.841	0.021 34	0.2761	0.9588	0.0675	2.178	−0.047 42
2	8	0.9815	0.0000	−0.1915	5.094	0.027 66	0.0000	1.0000	0.0000	1.963	−0.071 78
3	1	0.7887	−0.5929	−0.1624	3.806	0.005 10	0.6146	0.7653	0.1912	3.101	−0.006 26
3	2	0.8458	−0.5122	−0.1490	4.083	0.005 58	0.5329	0.8233	0.1951	2.829	−0.008 05
3	3	0.8823	−0.4492	−0.1408	4.281	0.006 65	0.4693	0.8625	0.1895	2.640	−0.010 78
3	4	0.9075	−0.3971	−0.1366	4.437	0.008 29	0.4167	0.8921	0.1747	2.498	−0.014 73
3	5	0.9293	−0.3439	−0.1344	4.588	0.010 51	0.3621	0.9199	0.1502	2.367	−0.020 37
3	6	0.9524	−0.2743	−0.1328	4.757	0.013 25	0.2896	0.9504	0.1138	2.225	−0.028 33
3	7	0.9769	−0.1682	−0.1315	4.938	0.016 21	0.1775	0.9821	0.0630	2.072	−0.038 64
3	8	0.9911	−0.0000	−0.1332	5.045	0.018 81	0.0000	1.0000	0.0000	1.982	−0.047 87
4	1	0.8257	−0.5460	−0.1415	3.976	0.003 97	0.5640	0.8031	0.1923	2.939	−0.005 38

4	2	0.8931	−0.4337	−0.1195	4.314	0.004 51	0.4495	0.8709	0.1986	2.607	−0.007 47
4	3	0.9276	−0.3578	−0.1069	4.510	0.005 61	0.3724	0.9080	0.1918	2.420	−0.010 45
4	4	0.9478	−0.3026	−0.1003	4.643	0.007 15	0.3162	0.9325	0.1746	2.302	−0.014 42
4	5	0.9629	−0.2519	−0.0965	4.757	0.009 08	0.2641	0.9532	0.1473	2.206	−0.019 58
4	6	0.9769	−0.1919	−0.0937	4.869	0.011 25	0.2016	0.9734	0.1089	2.115	−0.025 91
4	7	0.9897	−0.1102	−0.0917	4.972	0.013 32	0.1159	0.9915	0.0588	2.032	−0.032 60
4	8	0.9958	0.0000	−0.0914	5.021	0.014 57	0.0000	1.0000	0.0000	1.992	−0.036 73
5	1	0.8768	−0.4668	−0.1155	4.222	0.002 78	0.4809	0.8540	0.1985	2.699	−0.004 35
5	2	0.9410	−0.3274	−0.0862	4.562	0.003 53	0.3383	0.9186	0.2041	2.365	−0.006 81
5	3	0.9649	−0.2525	−0.0725	4.708	0.004 75	0.2620	0.9453	0.1944	2.229	−0.010 03
5	4	0.9765	−0.2052	−0.0660	4.795	0.006 30	0.2138	0.9612	0.1744	2.155	−0.014 02
5	5	0.9842	−0.1657	−0.0624	4.866	0.008 11	0.1732	0.9742	0.1448	2.101	−0.018 78
5	6	0.9907	−0.1224	−0.0599	4.932	0.010 01	0.1282	0.9861	0.1052	2.053	−0.024 04
5	7	0.9960	−0.0678	−0.0582	4.985	0.011 64	0.0711	0.9959	0.0559	2.014	−0.028 81
5	8	0.9983	0.0000	−0.0576	5.008	0.012 40	0.0000	1.0000	0.0000	1.997	−0.031 11
6	1	0.9485	−0.3079	−0.0737	4.592	0.001 72	0.3166	0.9247	0.2115	2.333	−0.003 38
6	2	0.9823	−0.1814	−0.0463	4.786	0.002 81	0.1872	0.9596	0.2102	2.145	−0.006 27
6	3	0.9906	−0.1317	−0.0369	4.849	0.004 18	0.1364	0.9710	0.1966	2.092	−0.009 70
6	4	0.9940	−0.1040	−0.0328	4.890	0.005 78	0.1081	0.9787	0.1744	2.063	−0.013 70
6	5	0.9961	−0.0824	−0.0306	4.928	0.007 55	0.0859	0.9859	0.1433	2.041	−0.018 23
6	6	0.9978	−0.0598	−0.0292	4.964	0.009 32	0.0625	0.9927	0.1030	2.021	−0.022 89
6	7	0.9991	−0.0325	−0.0283	4.991	0.010 75	0.0340	0.9979	0.0542	2.005	−0.026 75
6	8	0.9996	0.0000	−0.0280	5.002	0.011 34	0.0000	1.0000	0.0000	1.999	−0.028 37
7	1	1.0000	0.0000	0.0000	4.875	0.001 23	0.0000	0.9750	0.2221	2.051	−0.002 93
7	2	1.0000	0.0000	0.0000	4.885	0.002 54	0.0000	0.9770	0.2131	2.047	−0.006 06
7	3	1.0000	0.0000	0.0000	4.901	0.003 98	0.0000	0.9803	0.1975	2.040	−0.009 57
7	4	1.0000	0.0000	0.0000	4.923	0.005 60	0.0000	0.9847	0.1744	2.031	−0.013 58
7	5	1.0000	0.0000	0.0000	4.949	0.007 36	0.0000	0.9898	0.1428	2.021	−0.018 03
7	6	1.0000	0.0000	0.0000	4.974	0.009 10	0.0000	0.9948	0.1023	2.011	−0.022 50
7	7	1.0000	0.0000	0.0000	4.993	0.010 47	0.0000	0.9986	0.0537	2.003	−0.026 10
7	8	1.0000	0.0000	0.0000	5.000	0.011 02	0.0000	1.0000	0.0000	2.000	−0.027 55

The coordinates of the joints of the net of principal curvatures determined in this way (for the quarter of the net) are:

i	x_i	y_i	z_i
1	0	0	0
2	0	10.55	3.2969
3	0	14.75	4.6094
4	0	16.00	5.0000
8	0.25	0	0
9	5.88	6.55	1.7335
10	8.08	11.35	2.6059
11	8.80	16.00	2.9647
15	8.10	0	0
16	10.55	4.00	0.9740
17	12.35	10.38	2.1952
18	12.80	16.00	2.6365
22	17.50	0	0
23	17.50	1.70	0.3926
24	17.50	9.90	2.0266
25	17.50	16.00	2.5093

i	h	$l^0_{i,h}$	$f_{i,h}$				$s_{i,h}$	$l_{i,h}$
			x	y	s_1	s_2		
2	9	7.2814	2.940	8.550	3.9454	—	27.94	7.2679
9	16	5.3748	8.215	5.275	4.2781	—	27.48	5.3649
16	23	7.3437	14.055	2.850	4.5543	—	24.64	7.3317
3	10	8.9922	4.040	13.050	4.5734	—	23.03	8.9784
10	17	4.3980	10.215	10.865	4.8131	—	27.03	4.3901
17	24	5.1751	14.925	10.140	4.9287	—	33.12	5.1637
4	11	9.0323	4.400	16.000	5.1690	—	24.39	9.0176
11	18	4.0134	10.800	16.000	5.0158	—	26.02	4.0065
18	25	4.7017	15.150	16.000	5.0019	—	29.45	4.6925
8	9	8.8093	3.065	3.275	—	3.2234	20.40	8.7974
9	10	5.3517	6.980	8.950	—	2.4858	16.19	5.3460
10	11	4.7191	8.440	13.675	—	2.0775	13.73	4.7148
15	16	4.7907	9.325	2.000	—	2.9827	18.97	4.7847
16	17	6.7406	11.450	7.190	—	2.2542	12.56	6.7350
17	18	5.6552	12.575	13.190	—	2.0293	9.28	5.6517
22	23	1.7447	17.500	0.850	—	2.0533	15.08	1.7430
23	24	8.3612	17.500	5.800	—	2.0407	12.77	8.3541
24	25	6.1191	17.500	12.950	—	2.0072	9.91	6.1150

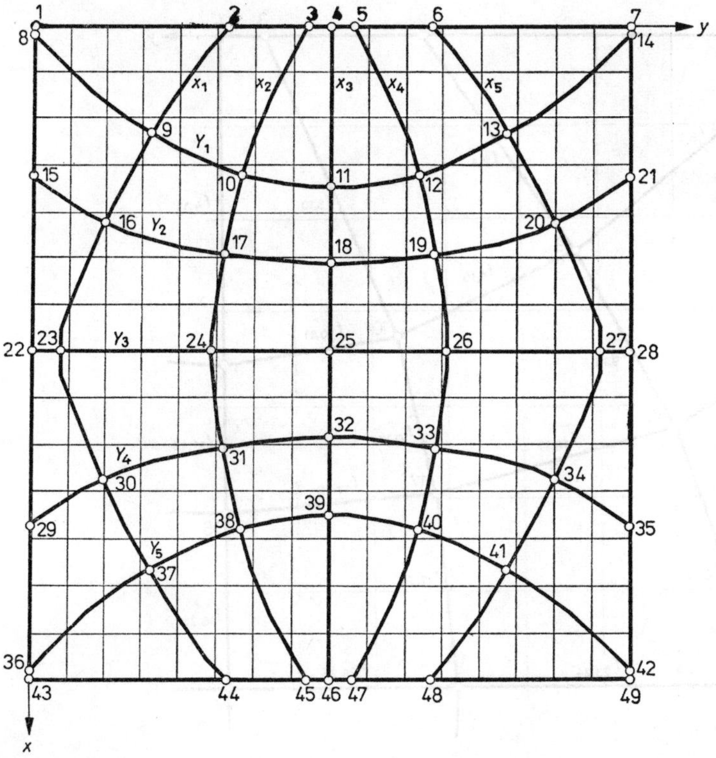

Fig. 5.5. Numerically constructed approximate direction lines of principal curvatures

One quarter of the net of principal curvatures is shown in Fig. 5.6. The approximate values of the cable forces were determined in such a manner that the values of the specific cable force intensity, s_1 and s_2 (belonging to the midpoint $f_{i,h}$ of the cable sections), were linearly interpolated with the aid of the values allocated to the adjacent rectangular net joints (if $f_{i,h}$ fell into an extreme field, it was extrapolated), and, then, this value of the intensity was multiplied by the average band width belonging to the cable section. The average band width was approximately determined by taking the average length of the four transverse cable sections joining the end points of the cable section (in some cases, as e.g. for cable sections 8,9, only two transverse cable sections join). By knowing the joint coordinates of the initial state, we can calculate the cable lengths $l_{i,h}^{(0)}$ while, with knowledge of the cable forces $s_{i,h}$, the unstressed lengths $l_{i,h}$ of the cable sections can also be determined. (In our example $EA=15$ MN for all cables.)
Once the values of the joint coordinates and the cable forces have been deter-

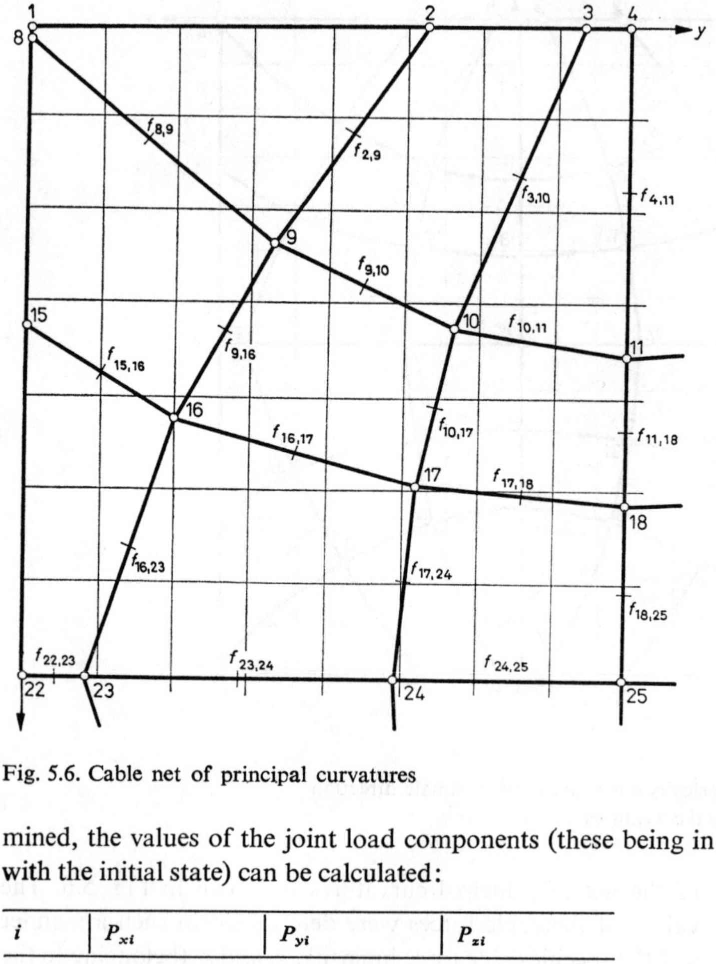

Fig. 5.6. Cable net of principal curvatures

mined, the values of the joint load components (these being in equilibrium with the initial state) can be calculated:

i	P_{xi}	P_{yi}	P_{zi}
9	5.0684	−1.6645	−0.7409
10	−0.9882	−1.7541	−1.0118
11	2.0190	0	−1.2803
16	6.9045	−1.3688	−0.3509
17	−4.1014	−0.2240	0.1062
18	−2.0293	0	0.1172
23	0	−13.2646	−3.0038
24	0	−3.4990	−0.4442
25	0	0	−0.0300

The sum of the vertical load components on the joints (in our example $\sum_{(i)} P_{zi} = -17.2418$ for the entire net) should equal the sum of the vertical

cable force components acting on the edge; this may serve as a check on the calculation.
If the sum of the absolute values of the unbalanced joint load components is taken as a guideline on the error, then 148.5 kN is obtained. This is reduced to 0.398 kN by the solution already obtained in the first step with the aid of the differential equation of the change of state. In the next step, the error is 0.756×10^{-3} kN, and in the third — concluding — step, it is 0.44×10^{-7} kN. The joint coordinates corresponding to this concluding step are:

i	x_i	y_i	z_i
9	5.879	6.558	1.7202
10	8.086	11.349	2.6251
11	8.800	16.000	2.9780
16	10.546	4.002	0.9688
17	12.352	10.380	2.1873
18	12.802	16.000	2.6293
23	17.500	1.702	0.3929
24	17.500	9.903	2.0161
25	17.500	16.000	2.5487

and the cable forces:

i	h	$S_{i,h}$	i	h	$S_{i,h}$
2	9	21.4682	8	9	20.2132
3	10	19.3794	9	10	17.6925
4	11	17.5726	10	11	13.8459
9	16	26.4293	15	16	17.2005
10	17	23.3025	16	17	11.8862
11	18	21.4254	17	18	8.4953
16	23	30.1688	22	23	30.8213
17	24	25.2425	23	24	11.3565
18	25	22.7005	24	25	6.5038

The example shown clearly illustrates how, owing to the rather rough approximation of the cable forces in the net of principal curvatures consisting of relatively few cables, the unbalanced load of the initial state was significant. In the course of restoring equilibrium, the cable forces changed considerably in some cable sections. Nevertheless, the joint coordinates did not undergo a change greater than 0.04 m compared to the initial state; thus the shape of the net — from the point of view of its design — did not change, for all practical purposes.

5.2. Net Connected to a Rigid Structure

By means of the calculation method outlined in Chapter 4 it is possible to determine the erection shape of the net fitting an arbitrary edge, and to calculate the values of the joint loads satisfying the conditions of equilibrium. If the net is fitted directly, or by means of flexible (and inherently kinematically indeterminate) edge cables, to a rigid edge or to rigid foundations, then the change of state related to the erection shape should be investigated according to what was said in Section 5.1. If, however, the net is connected either directly, or by means of intermediate flexible edge cables, to a rigid structure which behaves linearly, in a geometric sense, under the forces exerted on it by the net (i.e. if it suffers displacements that can be calculated with acceptable accuracy on the basis of the first-order theory), then it is superfluous to analyze the entire system on the basis of Eq. (5.7) valid for large displacements (that is, for geometrically non-linear structures). In such a case, it is more suitable to combine the equations valid for rigid structures with those valid for the net (including also the edge cables). It is the equation of the connected system, taking into consideration the interaction between the rigid structure behaving geometrically linearly and the geometrically non-linear net, which makes it possible to discuss the relationship between the loads and state characteristics of the entire connected system.

The method for the exact investigation of the connected system was elaborated by us for the case in which the rigid structure involved is an elastic bar structure behaving geometrically linearly (at least in the loading range considered). The methods of analysis of such bar structures were discussed in [25] in detail; hence the equation of the change of state based on the first-order theory is presented only briefly in Section 5.2.1.

The system of equations and the numerical procedure for the investigation of the change of state of the connected system (to be discussed in Section 5.2.2) enable us to attain an arbitrarily prescribed accuracy, provided the assumptions concerning the model are valid. A calculation with such a high degree of accuracy is rarely necessary. Instead, the determination of the state corresponding to the given loads with limited accuracy is frequently sufficient, with the supplementary requirement that it should be possible to calculate the exact value of the equilibrium load belonging to the state determined in this way and that this should not differ substantially from the prescribed load. A procedure of limited exactness meeting such requirements, considerably simpler than the one discussed in Section 5.2.2, is discussed in Section 5.3, and for the connected system, in Section 5.3.2.2.

Hereinafter it will always be assumed that the elastic bar structure (in short: R) and the cable net (in short: K) are connected in the erection state; therefore, at the moment of the connection the states of forces and of the geometry, of both partial systems, can be considered as known. Thus the dimensions and shapes of the bars and the cables belonging to the unloaded (unstressed) state can be determined.

5.2.1. Equation of the Change of State of the Elastic Bar Structure

Let us assume that the members of the bar structure to be discussed have straight axes, their rigidity is constant along the entire length of the bar, they are able to transmit tensile and compressive forces, as well as bending and twisting moments. (Our supplementary remarks concerning bars having spatially curved axes and non-uniform cross-sections are given in the Appendix). The bars are rigidly connected to each other and to the fixed foundations, i.e. there is no possibility of relative displacement.

This model neither excludes the consideration of bars having curvilinear axes, nor the taking into account of relative displacements at the individual bar connections. Namely, if it is necessary to use bars having curvilinear axes, then these can be replaced by bar polygons consisting of short bar pieces in such a manner that this approximation will not considerably influence the behaviour of the entire bar structure. Any relative displacement requirement at the bar connections can be satisfied by inserting a very short bar between the bar in question and the point of connection, the rigidity of which against the permitted relative movement is insignificantly small.

The equilibrium and displacement equations of the bar structure can be summarized in a single hypermatrix equation:

$$\begin{bmatrix} & \mathbf{G}_R^* \\ \mathbf{G}_R & \mathbf{F}_R \end{bmatrix} \begin{bmatrix} \mathbf{u}_R \\ \mathbf{s}_R \end{bmatrix} + \begin{bmatrix} \mathbf{q}_R \\ \mathbf{t}_R \end{bmatrix} = 0.$$

In this equation, $\mathbf{G}_R$ is a matrix containing the geometrical and topological characteristics of the bar structure, while $\mathbf{F}_R$ contains the elastic characteristics of the bars; $\mathbf{u}_R$ is the vector (column matrix) containing the displacement coordinates of the joints, $\mathbf{s}_R$ defines the bar forces, $\mathbf{q}_R$ the joint load components, $\mathbf{t}_R$ the relative displacements (independent of the joint loads) prescribed for each bar (the so-called kinematic loads, such as, e.g., the effect of temperature change).

Expounded in more detail, the structure of the submatrix of the equation is as follows.

The number of joints — excluding the points of support — is n, the number of bars is m (in accordance with the above stipulation, $n \leq m$, because otherwise the bar structure would be kinematically indeterminate).

The joints are marked by serial numbers, the bars by a double subscript formed with the serial number of the bar-end joints. Thus, e.g., j — joint subscript, j, k $(j<k)$ — bar subscript. In each row of the matrix $\mathbf{G}_R^*$ there are m matrices of order 6×6, and $\mathbf{G}_R^*$ consists of n such rows:

$$\mathbf{G}_R^* = [\mathbf{G}_{R\nu;\,j,k}^*]$$

$$\mathbf{G}_{R\nu;\,j,k}^* = \begin{cases} \mathbf{T}_{j,k}^*\mathbf{B}_{j,k}^*, & \text{if } \nu = j \\ -\mathbf{T}_{j,k}^*, & \text{if } \nu = k \\ 0, & \text{if } \nu \neq j \text{ and } \nu \neq k \end{cases}$$

$$\mathbf{T}_{j,k}^* = \begin{bmatrix} \mathbf{T}_{0;\,j,k}^* & \\ & \mathbf{T}_{0;\,j,k}^* \end{bmatrix}; \quad \mathbf{T}_{0;\,j,k}^* = [\mathbf{e}_{\xi;\,j,k}\mathbf{e}_{\eta;\,j,k}\mathbf{e}_{\zeta;\,j,k}]$$

$$\mathbf{B}_{j,k}^* = \begin{bmatrix} \mathbf{E} & 0 \\ \mathbf{B}_{0;\,j,k}^* & \mathbf{E} \end{bmatrix}; \quad \mathbf{B}_{0;\,j,k}^* = \begin{bmatrix} 0 & 0 & 0 \\ 0 & 0 & -l_{j,k} \\ 0 & l_{j,k} & 0 \end{bmatrix}.$$

$\mathbf{E}$ = unit matrix

$\mathbf{e}_{\xi;\,j,k}$, $\mathbf{e}_{\eta;\,j,k}$, $\mathbf{e}_{\zeta;\,j,k}$ are the unit vectors for the axes ξ, η, ζ of the local coordinate system allocated to bar j, k (Fig. 5.7), $l_{j,k}$ is the length of bar j, k. $\mathbf{F}_R = [\mathbf{F}_R, \mu, \nu; j, k]$ is an elasticity matrix (a so-called hyperdiagonal), of an $m\times m$ number of blocks of which only those where $\mu=j$, and $\nu=k$ differ from zero; thus the m number of blocks differing from zero can be expressed simply by the notation

$\mathbf{F}_{R_{j,k}}$,

where:

$$\mathbf{F}_{R_{j,k}} = \begin{bmatrix} \frac{l}{EA} & & & & & \\ & \frac{l^3}{3EI_\zeta} & & & & \frac{l^2}{2EI_\zeta} \\ & & \frac{l^3}{3EI_\eta} & & -\frac{l^2}{2EI_\eta} & \\ & & & \frac{l}{GI_\xi} & & \\ & & -\frac{l^2}{2EI_\eta} & & \frac{l}{EI_\eta} & \\ & \frac{l^2}{2EI_\zeta} & & & & \frac{l}{EI_\zeta} \end{bmatrix}_{j,k}.$$

The hypervector $\mathbf{u}_R$ of the joint displacement characteristics consists of a number n of vectors of dimension 6, each of which comprises the displace-

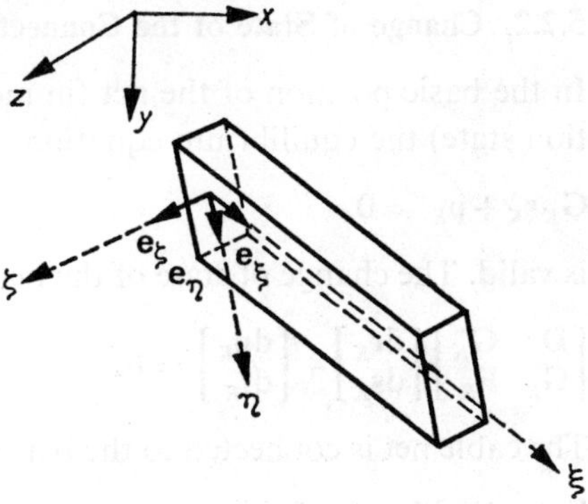

Fig. 5.7. Coordinate system allocated to an elastic bar

ment characteristics of one joint:

$$\mathbf{u}_{Rj} = \begin{bmatrix} v_{jx} \\ v_{jy} \\ v_{jz} \\ \varphi_{jx} \\ \varphi_{jy} \\ \varphi_{jz} \end{bmatrix} \begin{matrix} \left.\begin{matrix} \\ \\ \\ \end{matrix}\right\} \text{joint displacements} \\ \left.\begin{matrix} \\ \\ \\ \end{matrix}\right\} \text{joint axis-rotations} \end{matrix}$$

The hypervector $\mathbf{s}_R$ consists of a number m of vectors of dimension 6, the elements of which are equal to the index numbers of the force and couple components, respectively, acting on those ends of the individual bars which have a higher serial number:

$$\mathbf{s}_{Rj,k} = \begin{bmatrix} P_{j,k\xi} \\ P_{j,k\eta} \\ P_{j,k\zeta} \\ M_{j,k\xi} \\ M_{j,k\eta} \\ M_{j,k\zeta} \end{bmatrix} \begin{matrix} \left.\begin{matrix} \\ \\ \\ \end{matrix}\right\} \text{force components} \\ \left.\begin{matrix} \\ \\ \\ \end{matrix}\right\} \text{components of couples} \end{matrix}$$

The n number of vectors of dimension 6 in the hypervectors $\mathbf{q}_R$ and $\mathbf{t}_R$ contain the components of the loading forces and couples at the joint ($\mathbf{q}_{Rj}$) and the relative displacements along the bar (independent of the joint load), which are prescribed for each bar:

$$\mathbf{q}_{Rj} = \begin{bmatrix} R_{jx} \\ R_{jy} \\ R_{jz} \\ N_{jx} \\ N_{jy} \\ N_{jz} \end{bmatrix}; \quad \mathbf{t}_{Rj,k} = \begin{bmatrix} t_{j,k\xi} \\ t_{j,k\eta} \\ t_{j,k\zeta} \\ \vartheta_{j,k\xi} \\ \vartheta_{j,k\eta} \\ \vartheta_{j,k\zeta} \end{bmatrix}.$$

5.2.2. Change of State of the Connected System

In the basic position of the net (in most cases, this corresponds to the erection state) the equilibrium equation

$$\mathbf{G}_K\mathbf{s}_K+\mathbf{p}_K = 0$$

is valid. The change of state of the net is described by the equation

$$\begin{bmatrix} \mathbf{D} & \mathbf{G}_K^* \\ \mathbf{G}_K & \mathbf{F}_K \end{bmatrix}\begin{bmatrix} d\mathbf{r}_K \\ d\mathbf{s}_K \end{bmatrix}+\begin{bmatrix} d\mathbf{p}_K \\ d\mathbf{t}_K \end{bmatrix} = 0.$$

The cable net is connected to the bar structure, for which the equation

$$\begin{bmatrix} & \mathbf{G}_R^* \\ \mathbf{G}_R & \mathbf{F}_R \end{bmatrix}\begin{bmatrix} \mathbf{u}_R \\ \mathbf{s}_R \end{bmatrix}+\begin{bmatrix} \mathbf{q}_R \\ \mathbf{t}_R \end{bmatrix} = 0$$

is valid.

Let us assume that — after connection — the loads of the bar structure and the cable net change by a finitely small value, and let us also assume that the change of state corresponding to this can be described approximately by the equations

$$\begin{bmatrix} & \mathbf{G}_R^* \\ \mathbf{G}_R & \mathbf{F}_R \end{bmatrix}\begin{bmatrix} \Delta\mathbf{u}_R \\ \Delta\mathbf{s}_R \end{bmatrix}+\begin{bmatrix} \Delta\mathbf{q}_R \\ \Delta\mathbf{t}_R \end{bmatrix} = 0,$$

$$\begin{bmatrix} \mathbf{D} & \mathbf{G}_K^* \\ \mathbf{G}_K & \mathbf{F}_K \end{bmatrix}\begin{bmatrix} \Delta\mathbf{r}_K \\ \Delta\mathbf{s}_K \end{bmatrix}+\begin{bmatrix} \Delta\mathbf{p}_K \\ \Delta\mathbf{t}_K \end{bmatrix} = 0.$$

The connection of the two systems means that, at the points of connection, some displacement components in the two partial systems will be equal. If, in accordance with this, the displacement components which coincide during the connection with the displacement components of the other partial system in the course of every change of load and change of state, are provided in both partial systems with subscript I, while all the others are given the subscript II, then it is suitable to transcribe the above equations into the following from:

$$\begin{bmatrix} & & \mathbf{G}_{RI}^* \\ & & \mathbf{G}_{RII}^* \\ \mathbf{G}_{RI} & \mathbf{G}_{RII} & \mathbf{F}_R \end{bmatrix}\begin{bmatrix} \Delta\mathbf{u}_{RI} \\ \Delta\mathbf{u}_{RII} \\ \Delta\mathbf{s}_R \end{bmatrix}+\begin{bmatrix} \Delta\mathbf{q}_{RI} \\ \Delta\mathbf{q}_{RII} \\ \Delta\mathbf{t}_R \end{bmatrix} = 0,$$

$$\begin{bmatrix} \mathbf{D}_{11} & \mathbf{D}_{12} & \mathbf{G}_{KI}^* \\ \mathbf{D}_{21} & \mathbf{D}_{22} & \mathbf{G}_{RII}^* \\ \mathbf{G}_{KI} & \mathbf{G}_{RII} & \mathbf{F}_K \end{bmatrix}\begin{bmatrix} \Delta\mathbf{r}_{KI} \\ \Delta\mathbf{r}_{KII} \\ \Delta\mathbf{s}_K \end{bmatrix}+\begin{bmatrix} \Delta\mathbf{p}_{KI} \\ \Delta\mathbf{p}_{KII} \\ \Delta\mathbf{t}_K \end{bmatrix} = 0.$$

After connection, displacement increments $\Delta\mathbf{u}_{RI}$ and $\Delta\mathbf{r}_{KI}$ are identical; hence they may be given a common notation:

$$\Delta\mathbf{u}_{RI} = \Delta\mathbf{r}_{KI} = \Delta\mathbf{u}_{I}.$$

At the point of connection the sum of the joint loads (corresponding to the common displacement characteristics) will act, and it is not possible to show on which partial structure it acts because it is applied at their common joint. Thus, the load increment after connection is:

$$\Delta \mathbf{q}_{RI}+\Delta \mathbf{p}_{KI} \doteq \Delta \mathbf{q}_{I}.$$

Let us take out the first equation of each partial system and let us form their sum. The sum of equation

$$\mathbf{G}^*_{RI}\Delta \mathbf{s}_R+\Delta \mathbf{q}_{RI}=0$$

and equation

$$\mathbf{D}_{11}\Delta \mathbf{r}_{KI}+\mathbf{D}_{12}\Delta \mathbf{r}_{KII}+\mathbf{G}^*_{KI}\Delta \mathbf{s}_K+\Delta \mathbf{p}_{KI}=0$$

in accordance with the conditions of connection, is:

$$\mathbf{D}_{11}\Delta \mathbf{u}_{I}+\mathbf{D}_{12}\Delta \mathbf{r}_{KII}+\mathbf{G}^*_{KI}\Delta \mathbf{s}_K+\mathbf{G}^*_{RI}\Delta \mathbf{s}_R+\Delta \mathbf{q}_{I}=0.$$

The four remaining matrix equations of the two hypermatrix equations are included in a single hypermatrix equation by the equation just obtained. This will be the equation of the change of state of the connected system:

$$\left[\begin{array}{ccc:cc} \mathbf{D}_{11} & \mathbf{D}_{12} & 0 & \mathbf{G}^*_{KI} & \mathbf{G}^*_{RI} \\ \mathbf{D}_{21} & \mathbf{D}_{22} & 0 & \mathbf{G}^*_{KII} & 0 \\ 0 & 0 & 0 & 0 & \mathbf{G}^*_{RII} \\ \hdashline \mathbf{G}_{KI} & \mathbf{G}_{KII} & 0 & \mathbf{F}_K & 0 \\ \mathbf{G}_{RI} & 0 & \mathbf{G}_{RII} & 0 & \mathbf{F}_R \end{array}\right]\begin{bmatrix} \Delta \mathbf{u}_I \\ \Delta \mathbf{r}_{KII} \\ \Delta \mathbf{u}_{RII} \\ \Delta \mathbf{s}_K \\ \Delta \mathbf{s}_R \end{bmatrix}+\begin{bmatrix} \Delta \mathbf{q}_I \\ \Delta \mathbf{p}_{KII} \\ \Delta \mathbf{q}_{RII} \\ \Delta \mathbf{t}_K \\ \Delta \mathbf{t}_R \end{bmatrix}=0.$$

The coefficient matrix of the equation is symmetrical. Its simple build-up makes it possible to express blocks

$$\mathbf{A}_{ik}=\mathbf{A}^*_{ki}$$

(according to the partialization indicated by the dashed line) by the repeated application of the prescription of inverting

$$\begin{bmatrix} \mathbf{A} & \mathbf{B}^* \\ \mathbf{B} & \mathbf{C} \end{bmatrix}^{-1}=\begin{bmatrix} \mathbf{K}^{-1} & -\mathbf{K}^{-1}\mathbf{B}^*\mathbf{C}^{-1} \\ -\mathbf{C}^{-1}\mathbf{B}\mathbf{K}^{-1} & \mathbf{C}^{-1}+\mathbf{C}^{-1}\mathbf{B}\mathbf{K}^{-1}\mathbf{B}^*\mathbf{C}^{-1} \end{bmatrix}$$

$$\mathbf{K}=\mathbf{A}-\mathbf{B}^*\mathbf{C}^{-1}\mathbf{B}$$

with the aid of the blocks of the original coefficient matrix. The solution is:

$$\begin{bmatrix} \Delta \mathbf{u}_I \\ \Delta \mathbf{r}_{KII} \\ \Delta \mathbf{u}_{RII} \\ \Delta \mathbf{s}_K \\ \Delta \mathbf{s}_R \end{bmatrix}=\begin{bmatrix} \mathbf{A}_{11} & \mathbf{A}_{12} & \mathbf{A}_{13} & \mathbf{A}_{14} & \mathbf{A}_{15} \\ \mathbf{A}_{21} & \mathbf{A}_{22} & \mathbf{A}_{23} & \mathbf{A}_{24} & \mathbf{A}_{25} \\ \mathbf{A}_{31} & \mathbf{A}_{32} & \mathbf{A}_{33} & \mathbf{A}_{34} & \mathbf{A}_{35} \\ \mathbf{A}_{41} & \mathbf{A}_{42} & \mathbf{A}_{43} & \mathbf{A}_{44} & \mathbf{A}_{45} \\ \mathbf{A}_{51} & \mathbf{A}_{52} & \mathbf{A}_{53} & \mathbf{A}_{54} & \mathbf{A}_{55} \end{bmatrix}\begin{bmatrix} \Delta \mathbf{q}_I \\ \Delta \mathbf{p}_{KII} \\ \Delta \mathbf{q}_{RII} \\ \Delta \mathbf{t}_K \\ \Delta \mathbf{t}_R \end{bmatrix},$$

where

$$\mathbf{A}_{11} = (\mathbf{K}_{11} - \mathbf{K}_{13}\mathbf{K}_{33}^{-1}\mathbf{K}_{31} - \mathbf{K}_{12}\mathbf{K}_{22}^{-1}\mathbf{K}_{21})^{-1}$$

$$\mathbf{A}_{12} = -\mathbf{A}_{11}\mathbf{K}_{12}\mathbf{K}_{22}^{-1}$$

$$\mathbf{A}_{13} = -\mathbf{A}_{11}\mathbf{K}_{13}\mathbf{K}_{33}^{-1}$$

$$\mathbf{A}_{14} = -\mathbf{A}_{11}\mathbf{G}_{KI}^{*}\mathbf{F}_{K}^{-1} - \mathbf{A}_{12}\mathbf{G}_{KII}^{*}\mathbf{F}_{K}^{-1}$$

$$\mathbf{A}_{15} = -\mathbf{A}_{11}\mathbf{G}_{RI}^{*}\mathbf{F}_{R}^{-1} - \mathbf{A}_{13}\mathbf{G}_{RII}^{*}\mathbf{F}_{R}^{-1}$$

$$\mathbf{A}_{22} = \mathbf{K}_{22}^{-1} + \mathbf{K}_{22}^{-1}\mathbf{K}_{21}\mathbf{A}_{11}\mathbf{K}_{12}\mathbf{K}_{22}^{-1}$$

$$\mathbf{A}_{23} = -\mathbf{A}_{21}\mathbf{K}_{13}\mathbf{K}_{33}^{-1}$$

$$\mathbf{A}_{24} = \mathbf{A}_{21}\mathbf{G}_{KI}^{*}\mathbf{F}_{K}^{-1} - \mathbf{A}_{22}\mathbf{G}_{KII}^{*}\mathbf{F}_{K}^{-1}$$

$$\mathbf{A}_{25} = -\mathbf{A}_{21}\mathbf{G}_{RI}^{*}\mathbf{F}_{R}^{-1} - \mathbf{A}_{23}\mathbf{G}_{RII}^{*}\mathbf{F}_{R}^{-1}$$

$$\mathbf{A}_{33} = \mathbf{K}_{33}^{-1} + \mathbf{K}_{33}^{-1}\mathbf{K}_{31}\mathbf{A}_{11}\mathbf{K}_{13}\mathbf{K}_{33}^{-1}$$

$$\mathbf{A}_{34} = -\mathbf{A}_{31}\mathbf{G}_{RI}^{*}\mathbf{F}_{K}^{-1} - \mathbf{A}_{32}\mathbf{G}_{KII}^{*}\mathbf{F}_{K}^{-1}$$

$$\mathbf{A}_{35} = -\mathbf{A}_{31}\mathbf{G}_{RI}^{*}\mathbf{F}_{R}^{-1} - \mathbf{A}_{33}\mathbf{G}_{RII}^{*}\mathbf{F}_{R}^{-1}$$

$$\mathbf{A}_{44} = -\mathbf{F}_{K}^{-1} - \mathbf{F}_{K}^{-1}\mathbf{G}_{KI}\mathbf{A}_{14} - \mathbf{F}_{K}^{-1}\mathbf{G}_{KII}\mathbf{A}_{24}$$

$$\mathbf{A}_{45} = -\mathbf{F}_{K}^{-1}\mathbf{G}_{KI}\mathbf{A}_{15} - \mathbf{F}_{K}^{-1}\mathbf{G}_{KII}\mathbf{A}_{25}$$

$$\mathbf{A}_{55} = -\mathbf{F}_{R}^{-1} - \mathbf{F}_{R}^{-1}\mathbf{G}_{RI}\mathbf{A}_{15} - \mathbf{F}_{R}^{-1}\mathbf{G}_{RII}\mathbf{A}_{35}$$

Moreover $\mathbf{K}_{ik} = \mathbf{K}_{ki}^{*}$ and

$$\mathbf{K}_{11} = \mathbf{G}_{KI}^{*}\mathbf{F}_{K}^{-1}\mathbf{G}_{KI} + \mathbf{G}_{RI}^{*}\mathbf{F}_{R}^{-1}\mathbf{G}_{RI} - \mathbf{D}_{11}$$

$$\mathbf{K}_{12} = \mathbf{G}_{KI}^{*}\mathbf{F}_{K}^{-1}\mathbf{G}_{KII} - \mathbf{D}_{12}$$

$$\mathbf{K}_{13} = \mathbf{G}_{RI}^{*}\mathbf{F}_{R}^{-1}\mathbf{G}_{RII}$$

$$\mathbf{K}_{22} = \mathbf{G}_{KII}^{*}\mathbf{F}_{K}^{-1}\mathbf{G}_{KII} - \mathbf{D}_{22}$$

$$\mathbf{K}_{23} = 0$$

$$\mathbf{K}_{33} = \mathbf{G}_{RII}^{*}\mathbf{F}_{R}^{-1}\mathbf{G}_{RII}.$$

When the equation of the connected system was written, it was assumed that the load corresponding to the initial state changes by a small finite value. The load vector $\Delta\mathbf{q}$ and the kinematic load vector $\Delta\mathbf{t}$ are considered "small" if the procedure proposed below for the determination of the state corresponding to these is convergent. If this is not the case for the actual prescribed load increment, then two possibilities exist:

(a) for parts of the prescribed load, formed by appropriately chosen multiplying factors $\lambda (0 < \lambda < 1)$, the procedure is convergent; thus the required load can be applied in parts;

(b) the initial state of the connected system is the extreme position of the stable change of state; here the proposed procedure remains divergent even with $\lambda \to 0$.

Since we are not concerned here with case (b), it can be assumed that a load is involved for which the procedure is convergent.
It is assumed that the initial stresses ($\mathbf{s}_K$) and the joint coordinates of the cable net (and its geometrical matrix, $\mathbf{G}_K$, corresponding to this) are known. Of course, the data corresponding to the initial state of the bar structure are also assumed to be known. The loads compatible with the initial state can be determined by the expressions

$$\mathbf{q}_{\mathrm{I}} = -\mathbf{G}^*_{K\mathrm{I}}\mathbf{s}_K - \mathbf{G}^*_{R\mathrm{I}}\mathbf{s}_R$$

$$\mathbf{p}_{K\mathrm{II}} = -\mathbf{G}^*_{K\mathrm{II}}\mathbf{s}_K$$

$$\mathbf{q}_{R\mathrm{II}} = -\mathbf{G}^*_{R\mathrm{II}}\mathbf{s}_R.$$

Regarding the vector of the load increment

$$\Delta\mathbf{q} = \begin{bmatrix} \Delta\mathbf{q}_{\mathrm{I}} \\ \Delta\mathbf{p}_{K\mathrm{II}} \\ \Delta\mathbf{q}_{R\mathrm{II}} \\ \Delta\mathbf{t}_K \\ \Delta\mathbf{t}_R \end{bmatrix}$$

as given, it is possible to calculate the first approximation for the vector of the change of state:

$$\begin{bmatrix} \Delta\mathbf{u}_{\mathrm{I}}^{(1)} \\ \Delta\mathbf{r}_{K\mathrm{II}}^{(1)} \\ \Delta\mathbf{u}_{R\mathrm{II}}^{(1)} \\ \Delta\mathbf{s}_K^{(1)} \\ \Delta\mathbf{s}_R^{(1)} \end{bmatrix}.$$

With knowledge of the approximate value of the vector of the change of state, the corrected geometrical data of the cable net, as well as the stresses of the cable net and the bar structure, i.e. matrices

$$\mathbf{G}_{K\mathrm{I}}^{(1)};\quad \mathbf{G}_{K\mathrm{II}}^{(1)}$$

and vectors

$$\mathbf{s}_K^{(1)} = \mathbf{s}_K + \Delta\mathbf{s}_K^{(1)};\quad \mathbf{s}_R^{(1)} = \mathbf{s}_R + \Delta\mathbf{s}_R^{(1)}$$

are determined.

On the basis of this data it is now possible to determine the load vector compatible with the approximate value of the vector of the change of state:

$$\mathbf{q}_{\mathrm{I}}^{(1)} = -\mathbf{G}_{KI}^{(1)*}\mathbf{s}_{K}^{(1)} - \mathbf{G}_{RI}^{*}\mathbf{s}_{R}^{(1)}$$

$$\mathbf{p}_{K\mathrm{II}}^{(1)} = -\mathbf{G}_{K\mathrm{II}}^{(1)*}\mathbf{s}_{K}^{(1)}$$

$$\mathbf{q}_{K\mathrm{II}}^{(1)} = -\mathbf{G}_{R\mathrm{II}}^{*}\mathbf{s}_{R}^{(1)}.$$

The kinematic load vector compatible with the new state of the cable net is calculated with the aid of formula

$$\mathbf{t}_{K}^{(1)} = \mathbf{l}^{(1)} - \mathbf{l} - \mathbf{F}_{K}\Delta\mathbf{s}_{K}^{(1)}.$$

The elements of the error vector corresponding to the new state are:

$$\Delta\mathbf{q}_{\mathrm{I}}^{(1)} = \Delta\mathbf{q}_{\mathrm{I}} - (\mathbf{q}_{\mathrm{I}}^{(1)} - \mathbf{q}_{\mathrm{I}})$$

$$\Delta\mathbf{p}_{K\mathrm{II}}^{(1)} = \Delta\mathbf{p}_{K\mathrm{II}} - (\mathbf{p}_{K\mathrm{II}}^{(1)} - \mathbf{p}_{K\mathrm{II}})$$

$$\Delta\mathbf{q}_{R\mathrm{II}}^{(1)} = \Delta\mathbf{q} - (\mathbf{q}_{R\mathrm{II}}^{(1)} - \mathbf{q}_{R\mathrm{II}})$$

$$\Delta\mathbf{t}_{K}^{(1)} = \Delta\mathbf{t}_{K} - (\mathbf{t}_{K}^{(1)} - \mathbf{t}_{K}); \quad \mathbf{t}_{R}^{(1)} = 0.$$

If the error vector is smaller than the value prescribed in accordance with the accuracy of calculation, the computation is considered to be completed. If the error vector is still significant from the viewpoint of the prescribed accuracy, then, by regarding the new state as an initial state, the calculation is repeated taking the error vector as the load increment.

Example 5.3. Let us investigate the change of state of the connected system

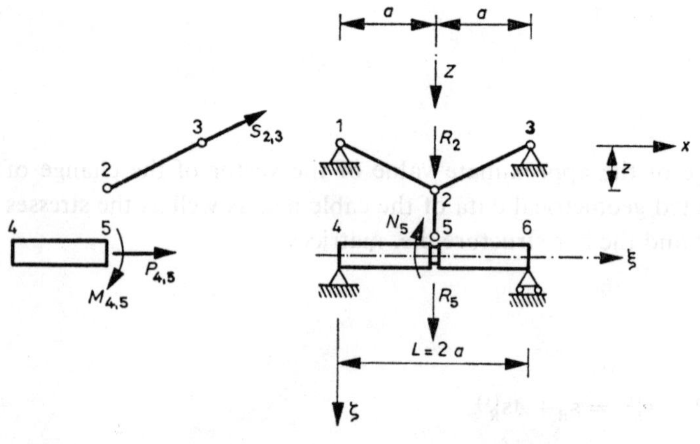

Fig. 5.8. Coupled system of three cables and a flexible bar

shown in Fig. 5.8. The data for the system is:

$a = 10.0$ m

$z = 1.0$ m

$E_K A = 80$ MN

$E_R I = 1.5$ MN m^2

$$\mathbf{G}^*_{KI} = \frac{z}{l}[-1 \quad -1]; \quad l = \sqrt{a^2+z^2} = l_{1,2} = l_{2,3}$$

$$\mathbf{G}_{KII} = 0; \quad \mathbf{s}^*_K = [s_{1,2} \quad s_{2,3}]$$

$$\mathbf{q}_K = 20 \text{ kN}; \quad \mathbf{q}_R = 0$$

$$\mathbf{F}_K = \frac{l}{E_K A}\begin{bmatrix} 1 & \\ & 1 \end{bmatrix}; \quad \mathbf{F}_R = \frac{a}{6E_R I}\begin{bmatrix} 2a^2 & 3a & & \\ 3a & 6 & & \\ & & 2a^2 & 3a \\ & & 3a & 6 \end{bmatrix},$$

$$\mathbf{G}_{RI} = \begin{bmatrix} -1 \\ 0 \\ 1 \\ 0 \end{bmatrix}; \quad \mathbf{G}_{RII} = \begin{bmatrix} a & \\ 1 & -1 \\ & a \\ & 1 \quad -1 \end{bmatrix}$$

$$\mathbf{D}_{11} = -\frac{a^2}{l^3}(s_{1,2}+s_{2,3}); \quad \mathbf{D}_{12} = \mathbf{D}_{22} = 0.$$

The load increment of the connected system is:

$$\Delta\mathbf{q}_I = 20 \text{ kN}; \quad \Delta\mathbf{t}_K = \Delta\mathbf{t}_R = 0.$$

Let us consider the iteration process to be completed at the ith step if the guideline (or norm) for the error vector is:

$$\|\Delta\mathbf{q}_I^{(i)}\| \leqslant \varepsilon_q = \|\mathbf{q}_K+\Delta\mathbf{q}_I\| \times 10^{-3},$$

and

$$\|\Delta\mathbf{t}_K^{(i)}\| \leqslant \varepsilon_t = 0.01 \text{ mm},$$

where now the quantity $\|\mathbf{a}\| = \sqrt{\mathbf{a}^*\mathbf{a}}$ is called the norm of vector $\mathbf{a}$.
The approximate value of the vector of the change of state is:

$$\Delta\mathbf{u}_I^{(1)} = 0.107\,278 \text{ m}$$

$$\Delta\mathbf{s}_K^{(1)} = 84.972\,619\begin{bmatrix} 1 \\ 1 \end{bmatrix} \text{kN}$$

and the new load increment necessary to achieve the prescribed load increment:

$$\Delta \mathbf{q}_{\mathrm{I}}^{(1)} = -1.789\,670 \text{ kN}$$

$$\Delta \mathbf{t}_{K}^{(1)} = -0.566 \times 10^{-3} \begin{bmatrix} 1 \\ 1 \end{bmatrix} \text{m}.$$

The increment of the vector of the change of state corresponding to this is given by

$$\Delta \mathbf{u}_{\mathrm{I}}^{(2)} = -0.011\,682 \text{ m}$$

$$\Delta \mathbf{s}_{K}^{(2)} = -5.720\,077 \begin{bmatrix} 1 \\ 1 \end{bmatrix} \text{kN},$$

while the new load increment necessary to achieve the prescribed load increment (the error vector) becomes:

$$\Delta \mathbf{q}_{\mathrm{I}}^{(2)} = -0.013 \text{ kN}$$

$$\Delta \mathbf{t}_{K}^{(2)} = -6.7 \times 10^{-6} \begin{bmatrix} 1 \\ 1 \end{bmatrix} \text{m}.$$

Thus, according to the prescription concerning the error vector, the calculation can be considered to be completed because

$$\|\Delta \mathbf{q}_{\mathrm{I}}^{(2)}\| = 0.013 \text{ kN} < \varepsilon_q = 0.04 \text{ kN}$$

and

$$\|\Delta \mathbf{t}_{K}^{(2)}\| = 0.0095 \text{ mm} < \varepsilon_t = 0.01 \text{ mm}.$$

5.3. A Method of Limited Accuracy for the Investigation of a Rectangular Cable Net Stretched on a Rigid Edge

In the Introduction and in Chapters 1 and 4 the attention of the reader was drawn to the advantages arising from the application of the rectangular cable net. In Chapter 4 it was indicated that the erection shape of the rectangular cable net — in the case of an arbitrary rigid or flexible edge — can be determined to any required accuracy. If, however, a change from the erection load takes place on the net fixed to the edge, then the cable forces and the joint coordinates change, and the ground plan of the cables connected to each other during erection will no longer be a set of straight lines intersecting one another at right angles.

The variation of the cable forces causes displacements on the rigid or flexible edge, and because of this further net deformations occur.

Taking all these factors into consideration, the exact calculation of the change of state of the cable net connected to a rigid structure is possible only in the way discussed in Sections 5.1. and 5.2. However, the accuracy obtainable by the procedure described there is seldom necessary. Nevertheless, it may be justified to develop a calculation method with limited accuracy — a "golden mean" between the exact procedure discussed in Sections 5.1 and 5.2 and the approximate procedure described in Chapter 2 — the criteria of which can be characterized by the following:

1. its basis is the exact calculation of the approximate model of the real net;
2. it enables an approximately exact determination of the horizontal displacements of the net joints;
3. it makes it possible to consider the elastic displacements of the solid edge;
4. with knowledge of the approximately exact geometrical data and cable forces of the net, it makes it possible to determine exactly the loads which are in equilibrium with these and the deviation from the prescribed load;
5. compared to the exact procedure, it significantly reduces the amount of calculation work, it can be programmed easily and it does not make a particularly high storage demand on the computer.

The model on which the method is based differs from the assumptions presented in Section 1.2.2.1 in that

1. it consists of a finite number of cables; thus it cannot be considered to be a continuum;
2. it takes the elastic, horizontal displacements of the rigid edge into consideration;
3. although it excludes the horizontal displacements of the net joints (completely neglected by other methods), it offers the possibility for their approximate determination once the vertical displacements are known.

The latter property of the model can be realized *in principle* in such a manner that the intersecting cables are made frictionless, and that it is ensured, by suitable guide rails, that the cables should not deviate from their vertical plane in the course of the change of state.
Only vertical loads can act on the net joints; therefore the horizontal components of the cable forces will be identical in every section of a given cable. Thus, the advantage of the calculation of the model following from the foregoing manifests itself in the fact that, with coordinates z corresponding to the state of equilibrium,

1. the horizontal components of the cable forces belonging to each cable will be the unknown quantity, and
2. it is necessary to write only one compatibility equation per cable.

If the net has a number m of cables in both directions (thus a number $2m$ in all), then contrary to the operation of an m^6 order of magnitude, necessary in the case of the exact procedure, only an operation of an m^5 order of magnitude (or in the case of the application of the method discussed in Section 4.4, only of an m^4 order of magnitude) is necessary if the procedure with limited accuracy is applied.
The vertical displacements of the net calculated on the basis of the model just outlined can be determined with arbitrarily prescribed accuracy, according to the procedure described below. On the other hand, the procedure can determine the horizontal displacement only with limited accuracy because it assumes, during the calculation, that the vertical displacements of the joints do not change when the horizontal displacements take place. Expanding on this: when the erection shape is achieved, the intersecting cables are definitely connected, and thus the joints suffer both horizontal and vertical displacements under the effect of the surplus load; in the simplified model it is assumed that the values of the vertical displacements are equal to those of the vertical displacements determined with the exact method. As a result of the investigation of the change of state, the values of both, the cable forces between two joints in the cables of the net and of the displacement coordinates of the joint, are obtained with limited accuracy. On the other hand, the reliability of the final result of the procedure with limited accuracy can be measured exactly with the aid of the error vector discussed in Sections 5.1.4 and 5.2.2.

5.3.1. Cable Net Equations of General Validity

The equilibrium and compatibility equations of the cable nets are investigated separately because the numerical treatment of the non-linear relationships between the loads, internal forces and displacement characteristics can be more easily obtained in this way. The relationships to be discussed are generated by an approach slightly different from previous ones, partly to show an alternative way of tackling the problem, partly in order to simplify the resulting relationships as much as possible.
The equilibrium equations of the cable net express that the sum of the forces acting on the individual inner and edge joints, is zero.
The equilibrium equations concerning the inner joints comprise the coordi-

nates of the inner joints, the external forces acting on the joints and the cable force components parallel to the reference plane (i.e. ground plan). (If the number of the inner joints is N and the number of all the cable members is M, then the number of the equilibrium equations that can be written is $3N$, while the number of the unknown quantities they contain is $M+3N$ (i.e. 3–3 coordinates per joint, and the cable force or the horizontal component of the cable force for each cable member). Namely, the definition of the cable net makes it possible that the cable force component, parallel to the reference plane (hereinafter force H for short) should be different, in absolute value, for each cable member. However, the forces arising in each cable member are already uniquely determined by the coordinates of the joints and by the forces H. The magnitude of the internal forces acting on the joints depends on the change in the forces H (M number in all) and the joint coordinates ($3N$ number in all); their resultant must equilibrate, at each joint, the load acting at that joint, that is, the resultant of the internal forces and the 3–3 linearly independent components of the load must each satisfy the requirement of one zero-valency. Thus, the $M+3N$ number of unknown quantities have to satisfy the requirements expressed by the $3N$ number of equations in all, so that the $3N$ number of equilibrium equations will contain a number M of free parameters. In order to describe the geometrical position and the state of equilibrium of the net (referred to simply as "the state of the net") uniquely, compatibility conditions must be prescribed for the M cables of the net.

The compatibility equations of the net express the fact that the length of the cable section between every two joints will differ from its initial length (which is the length of the unstressed cable between the two adjacent joints, having an initial temperature) by an amount equal to its elongation (or shortening as the case may be) in consequence of the tensile force and the temperature change arising in the cable. These equations will also contain the forces H and the joint coordinates as the unknown quantities.

Thus, the two equation systems (the equilibrium equations of the inner joints and the compatibility equations of the cable sections) consist of a number of $M+3N$ equations while the number of unknown quantities is $M+3N$. As concerns the edge, the equilibrium and compatibility equations must be written only if the edge is not absolutely rigid. In the case of a rigid edge, the relevant relations consist of equilibrium and deformation equations for the edge girder — generally with a spatially curved axis. We shall briefly revert to their discussion later.

The cable net described above is entirely general, but the equations that describe it do not allow the formulation of an especially advantageous algorithm. Moreover, they are almost unsuitable for calculation purposes because the relationships between the $M+3N$ number of unknowns contained in them are given partly by irrational functions. But, although the $M+3N$ number of unknown quantities cannot, in general, be calculated directly from these equations, one must not think that writing them in a general form is completely useless for practical purposes. Namely, it will be seen from our later discussions that, with given net joint coordinates and with the prescription of the components of the cable forces parallel to the reference plane, it is advantageous to use the above equations in calculating those external forces acting on the joints under which equilibrium is satisfied, and those cable temperatures for which the compatibility conditions are met. This fact justifies our writing of the equilibrium and compatibility equations for the general net prior to the discussion of the special nets. The jth joint of the cable net with the cable sections belonging to it is shown in Fig. 5.9.

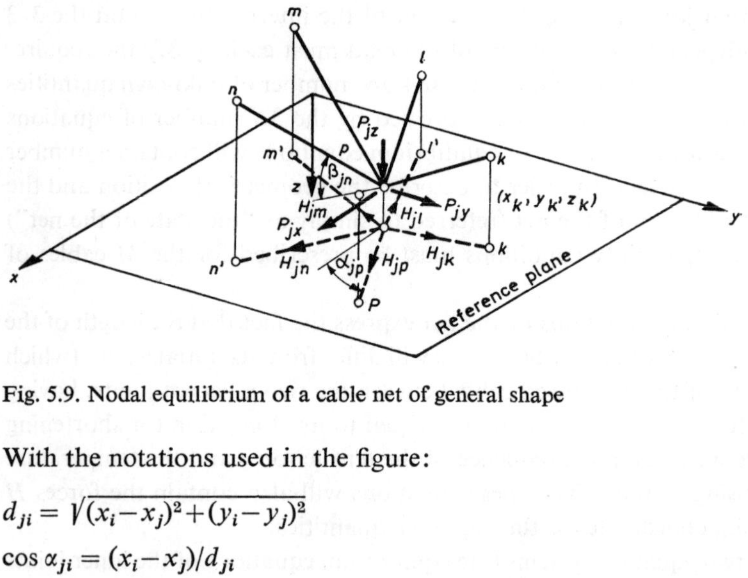

Fig. 5.9. Nodal equilibrium of a cable net of general shape

With the notations used in the figure:

$$d_{ji} = \sqrt{(x_i-x_j)^2+(y_i-y_j)^2}$$

$$\cos\alpha_{ji} = (x_i-x_j)/d_{ji}$$

$$\sin\alpha_{ji} = (y_i-y_j)/d_{ji}$$

$$\tan\beta_{ji} = (z_i-z_j)/d_{ji}; \quad (i = k, l, m, n, p),$$

the equilibrium equations of the jth joint are:

$$\begin{aligned} &\sum_{(i)} H_{ji}\cos\alpha_{ji}+P_{jx} = 0 \\ &\sum_{(i)} H_{ji}\sin\alpha_{ji}+P_{jy} = 0 \\ &\sum_{(i)} H_{ji}\tan\beta_{ji}+P_{jz} = 0 \quad (i = k, l, m, n, p). \end{aligned} \tag{5.13}$$

The magnitude of the cable force (j, k) is:

$$s_{ji} = H_{ji}\frac{1}{\cos\beta_{ji}},$$

where

$$\cos\beta_{ji} = \frac{d_{ji}}{\sqrt{(x_i-x_j)^2+(y_i-y_j)^2+(z_i-z_j)^2}}.$$

Let us indicate the length of the unloaded cable section (j, i) at temperature $t=0$ by l_{ji}. The compatibility involving the cable length between the end-point coordinates of cable section (j, i), the cable force with component H_{ji}, as well as the elongation caused by the cable temperature t_{ji} is expressed by the following equation:

$$d_{ji}\frac{1}{\cos\beta_{ji}} = l_{ji}+\frac{H_{ji}}{EA_{ji}}d_{ji}\frac{1}{\cos^2\beta_{ji}}+t_{ji}\alpha_t d_{ji}\frac{1}{\cos\beta_{ji}}, \tag{5.14}$$

where A_{ji} is the cross-sectional area of the cable, E is the modulus of elasticity of the cable material and α_t is the linear coefficient of thermal expansion of the cable. From Eqs (5.13)–(5.14) — as indicated — with given joint coordinates and force components H, the joint forces P and the cable temperatures ensuring compatibility can be uniquely calculated.

We can already observe here the form of Eqs (5.13)–(5.14), very characteristic for cable nets. P_{jx}, P_{jy}, P_{jz} and t_{ji} can be explicitly expressed, by means of these equations, as functions of the joint coordinates and the forces H:

$$P_{jx} = -\sum_{(i)} H_{ji}\cos\alpha_{ji}$$

$$P_{jy} = -\sum_{(i)} H_{ji}\sin\alpha_{ji}$$

$$P_{jz} = -\sum_{(i)} H_{ji}\tan\beta_{ji}$$

$$t_{ji} = \left(1-\frac{H_{ji}}{EA_{ji}}\frac{1}{\cos\beta_{ji}}-\frac{l_{ji}}{d_{ji}}\cos\beta_{ji}\right)\frac{1}{\alpha_t}.$$

Namely, it must not be forgotten that EA_{ji}, l_{ji} and α_t are constants given in advance, and α_{ji}, β_{ji}, and d_{ji} are values depending only on the joint coordinates. The $3N+M$ number of variables P_{jx}, P_{jy}, P_{jz} and t_{ji} — the state characteristics of the net — are *unique* functions of the $3N+M$ number of independent variables x_j, y_j and H_{ji} and hence can be obtained directly. Let us include these two groups of variables into two *vectors* of dimension $3N+M$, in the following manner:

$$\mathbf{v} = \begin{bmatrix}\mathbf{t}\\ \mathbf{P}\end{bmatrix}; \quad \mathbf{r} = \begin{bmatrix}\mathbf{H}\\ \mathbf{z}\end{bmatrix},$$

where

$$\mathbf{t} = \begin{bmatrix} t_{12} \\ t_{13} \\ \vdots \\ t_{jk} \\ \vdots \end{bmatrix} \begin{matrix} (1 \\ (2 \\ \\ \\ (M \end{matrix} ; \quad \mathbf{P} = \begin{bmatrix} P_{1x} \\ \vdots \\ P_{Nx} \\ P_{1y} \\ \vdots \\ P_{Ny} \\ P_{1z} \\ \vdots \\ P_{Nz} \end{bmatrix} \begin{matrix} (1 \\ \\ (N \\ (N+1 \\ \\ (2N \\ (2N+1 \\ \\ (3N \end{matrix}$$

$$\mathbf{H} = \begin{bmatrix} H_{12} \\ H_{13} \\ \vdots \\ H_{jk} \\ \vdots \end{bmatrix} \begin{matrix} (1 \\ (2 \\ \\ \\ (M \end{matrix} ; \quad \mathbf{z} = \begin{bmatrix} x_1 \\ x_2 \\ \vdots \\ x_N \\ y_1 \\ y_2 \\ \vdots \\ y_N \\ z_1 \\ z_2 \\ \vdots \\ z_N \end{bmatrix} \begin{matrix} (1 \\ (2 \\ \\ \\ \\ \\ \\ \\ \\ \\ \\ (3N \end{matrix} .$$

The relationship between the two variables (**v** and **r**), according to the above, can be written in the form

$$\mathbf{v} = \mathbf{v}(\mathbf{r})$$

as well. The relationship is unique and explicit, i.e. *a single* vector **v** belongs to every vector **r**, on condition that the basic requirement concerning the net (i.e. there must not be a vertical cable section in it)

$$|x_i - x_j| + |y_i - y_j| \neq 0, \quad \text{if} \quad i \neq j$$

be fulfilled and **r** be finite.

The drawback of the vector function $\mathbf{v}=\mathbf{v}(\mathbf{r})$ is that — apart from quite simple cases — the required values cannot be explicitly obtained from it. Namely, in general, the vector **v** (the elements of which are the given joint load components and the cable temperatures) should be regarded as given, and vector **r** (the elements of which are the joint coordinates and forces H corresponding to the state of equilibrium) should be regarded as the unknown. The inversion of the function $\mathbf{v}=\mathbf{v}(\mathbf{r})$, i.e. the generation of a function $\mathbf{r}=\mathbf{r}(\mathbf{v})$, is generally not possible due to the intricate relations involved;

therefore, the determination of **r** corresponding to a given **v** is possible only with the aid of successively approximating procedures. The problem is made difficult by the fact that the function $\mathbf{v}=\mathbf{v}(\mathbf{r})$ is non-linear, and the approximations resting on a linear basis may be considered to be valid only for a small change of state of the cable net.

The *basic state*, which can be regarded as having been determined in some way or other, plays a decisive role in the determination of **r** belonging to the prescribed **v**. (This determination can be made, e.g., by the calculation of a quite arbitrarily adopted vector **r** and a vector **v** uniquely belonging to this). The state belonging to the prescribed **v** can be found by successive approximations, starting from the basic state. If the effect of the infinitesimally small change taking place from the basic state is known, i.e. if one can speak of the relationship

$$d\mathbf{v} = \mathbf{D}\ d\mathbf{r},$$

then, by means of the expression $d\mathbf{r}=\mathbf{D}^{-1}\,d\mathbf{v}$ it is possible to bring about such a change (related to the basic state given by the vectors $\mathbf{r}_0$, $\mathbf{v}_0$) in which $\mathbf{v}_0+d\mathbf{v}$ is now "nearer" to the state characteristic $\mathbf{v}_1$ to be achieved. If one has really got "nearer" to the state to be achieved in this way, then in the following step this state must be considered to be the basic state.

5.3.2. The Rectangular Cable Net

The so-called rectangular cable net, having a projection on the reference plane that consists of two sets of straight lines intersecting each other at right angles, plays an important role among the cable net structures used in practice. If the distance between two adjacent parallel straight lines is constant in the ground plan of the rectangular cable net, then a *uniform rectangular cable net* is involved. We shall now treat this net model in detail.

The procedure for the basic model is summarized as follows (Fig. 5.10).

1. The projection of the edge line relating to plane (x, y) is a rectangle, with side lengths $(m+1)a$ and $(n+1)b$, parallel to the x and y axes, respectively. The centre line of the rigid edge is a closed space-curve.
2. The net consists of a number n of cables lying in the planes parallel to the plane (x, z) and a number m of cables lying in the planes parallel to the plane (y, z), rigidly fixed to the edge. The lengths of these cables are (when unload-

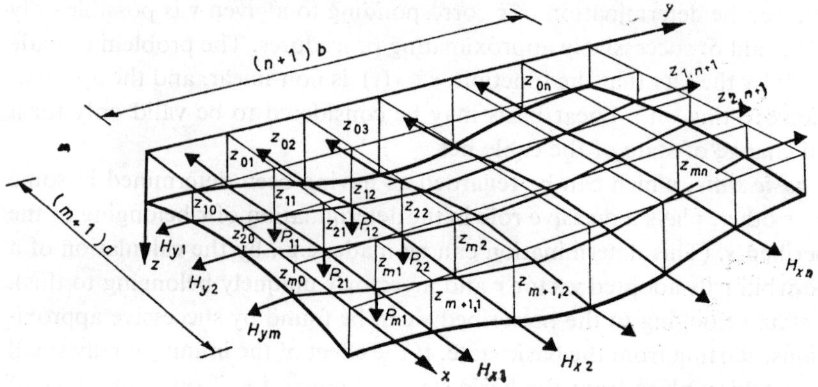

Fig. 5.10. Nodal loads and heights of a rectangular cable net

ed and at the initial temperature, i.e. $H=0$, $t=0$) l_{xk} and l_{yj} ($k=1, 2, \ldots, n$; $j=1, 2, \ldots m$), respectively.

3. The coordinates $x_k=ka$, $y_j=jb$ of the joints of the intersecting cables remain unchanged during the change of state, and the coordinates $z_{j,k}$ of the intersecting cables are identical at joint (j, k). (Thus, the same points of the cables will not always get into the joint in the course of the change of state!).

4. Only external forces $P_{j,k}$ parallel to the z axis act on the joints; therefore (and in connection with 3) the components of the cable forces parallel to the plane (x, y) are constant in each cable and their values are X_{xk} and H_{yj}, respectively.

5. In each cable, and for each state of the net, the same temperature prevails along the cable: t_{xk}, and t_{yj}, respectively. The cross-section of each cable (A_{xk} and A_{yj}, respectively) is constant. The modulus of elasticity E and the coefficient of thermal expansion α_t of each cable are identical.

One of the significant advantages of the method (and the model serving as a basis for it) is that one compatibility equation must be written for each cable parallel to planes (x, z) and (y, z), respectively (and not one for each cable section), viz. either of the following two equations applies:

$$l_{xk} = a \sum_{j=1}^{m+1} \frac{1}{\cos \alpha_{j,k}} - a t_{xk} \alpha_t \sum_{j=1}^{m+1} \frac{1}{\cos \alpha_{j,k}} - \frac{H_{xk} a}{E F_{xk}} \sum_{j=1}^{m+1} \frac{1}{\cos^2 \alpha_{j,k}}$$

$$l_{yj} = b \sum_{k=1}^{n+1} \frac{1}{\cos \beta_{j,k}} - b t_{yj} \alpha_t \sum_{k=1}^{n+1} \frac{1}{\cos \beta_{j,k}} - \frac{H_{yj} b}{E F_{yj}} \sum_{k=1}^{n+1} \frac{1}{\cos^2 \beta_{j,k}}, \qquad (5.15)$$

where

$$\frac{1}{\cos \alpha_{j,k}} = \sqrt{1+\left(\frac{z_{j,k}-z_{j-1,k}}{a}\right)^2}; \quad \frac{1}{\cos \beta_{j,k}} = \sqrt{1+\left(\frac{z_{j,k}-z_{j,k-1}}{b}\right)^2},$$

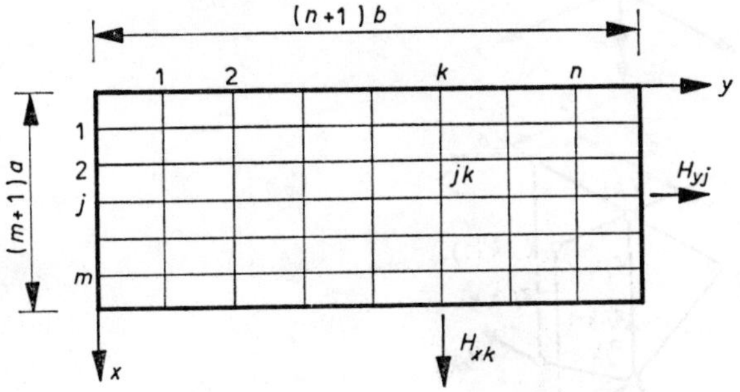

Fig. 5.11. Ground plan of a rectangular cable net

and l_{xk} and l_{yj} denote the original length (i.e. in the unloaded state and measured at the initial temperature) of the kth and jth cables, respectively (see, in ground plan, in Fig. 5.11) parallel to planes (x, z) and (y, z), respectively. (Of course, in this case the meanings of α and β, and that of the subscripts differ from the notation used in Section 5.3.1.)

The compatibility equations express the fact that the lengths of the cables, determined by the z_{jk} coordinates characterizing the state, may only differ from the basic lengths l_{xj} and l_{yj} by the elongation due to the cable forces and the temperature change.

The equilibrium of the joints of the net is expressed by the equations

$$\frac{1}{a} H_{xk}(-z_{j-1,k}+2z_{j,k}-z_{j+1,k})+\frac{1}{b} H_{yj}(-z_{j,k-1}+2z_{j,k}-z_{j,k+1}) = P_{j,k} \tag{5.16}$$

$(j=1, 2, \ldots, m;\ k=1, 2, \ldots, n)$. Their meaning becomes clear by reference to Fig. 5.12.

Equation (5.16) is valid for all the mn inner joints of the net. By a suitable grouping of the quantities contained in it, we can set up the Poisson-type differential equation system of the basic rectangular net.

In the equation system, H, P, a and b are constants; furthermore, the edge-point heights are given. The problem consists in calculating the z coordinates of the inner joints. The equation system itself is a simple band matrix system, where the band width exceeds by one twice the maximum number of the unidirectional cables. (The computers in use have ready programs for the solution of such systems of equations.)

Essentially, the same equation system is also valid when the edge of the cable net is not a rectangle in ground plan.

The coefficient matrix of the equation system can easily be written if the edge is as shown in Fig. 5.13, i.e., if the edge sections differing from the rectan-

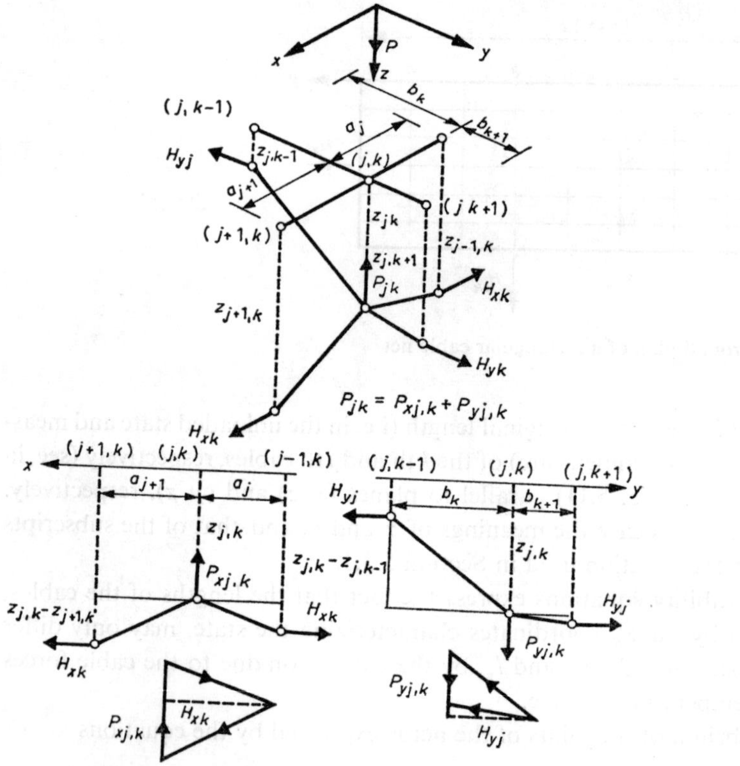

Fig. 5.12. Nodal equilibrium of a rectangular cable net

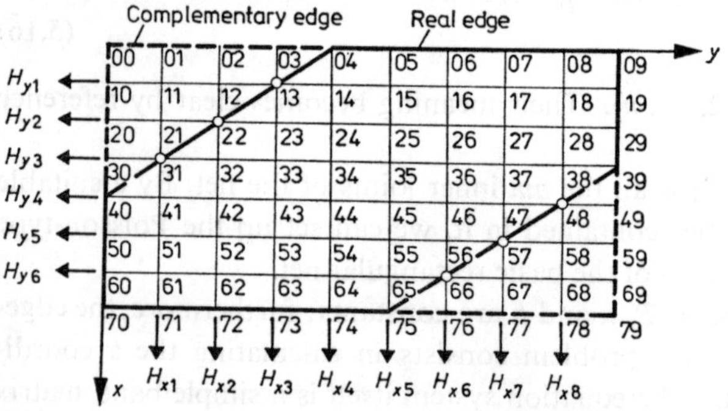

Fig. 5.13. Cable net streched on a non-rectangular edge

gular ground plan pass through net joints. In this case too, the equation system (5.16) is written only for the inner joints (i.e. those within the real edge), but in the equations for the points in the vicinity of the edge, the z coordinates of the edge points are considered as known quantities. (E.g. in Fig. 5.13, z_{22} should be regarded as known in the equation for point 23.)

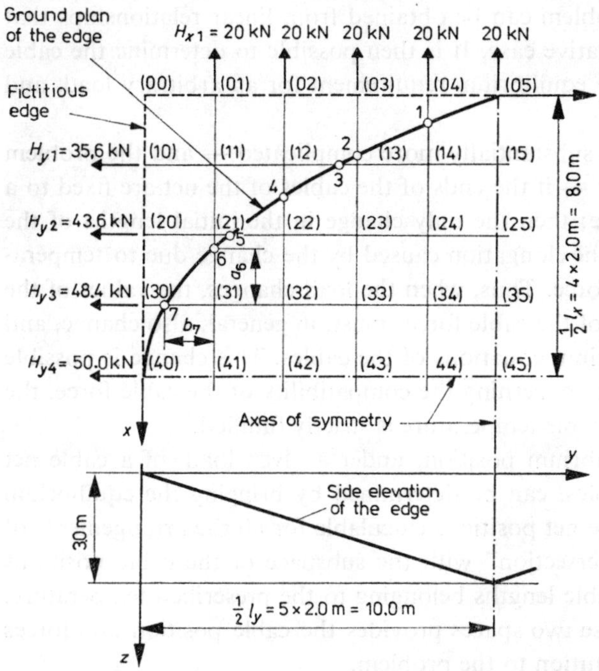

Fig. 5.14. Cable net stretched on an edge curved in ground plan

If the edge is similar to that shown in Fig. 5.14, then the only complication in setting up the equations is that, in the equations for the points in the vicinity of the edge, the actual ground plan distances must be used.
For example, on the basis of (5.16), the equation of joint 31 in Fig. 5.14 will have the following form:

$$H_{x1}\left(\frac{1}{a_6}+\frac{1}{a}\right)z_{31}-H_{x1}\frac{1}{a}z_{41}+H_{y3}\left(\frac{1}{b_7}+\frac{1}{b}\right)z_{31}-H_{y3}\frac{1}{b}z_{32}=$$

$$=P_{31}+H_{x1}\frac{1}{a_6}z_6+H_{y3}\frac{1}{b_7}z_7,$$

(the right side of the equation contains heights z_6, z_7 prescribed for edge points 6 and 7).

5.3.2.1. Net Fixed to a Rigid Edge

It can be seen, on the basis of the foregoing, that the equilibrium of the equidistantly arranged, rectangular cable net can be investigated without any difficulty provided the value of the horizontal projection of the cable force (the so-called force H) is a prescribed constant for each cable. In such a case, the solution of the problem can be obtained from linear relationships that can be treated with relative ease. It is then possible to determine the cable positions satisfying the equilibrium requirement for an arbitrary load, and vice versa.

The situation becomes substantially more complicated — and the problem also ceases to be linear — if the ends of the cables of the net are fixed to a rigid or an elastic edge; then the only change in the initial lengths of the cables corresponds to the elongation caused by the change due to temperature and/or the cable force. Thus, when the load changes, the values of the horizontal projections of the cable forces must, in general, also change, and this affects the equilibrium equations of the cables. This change is possible only if the requirement concerning the compatibility of the cable force, the cable position and the cable temperature is strictly fulfilled.

In principle, the equilibrium position, under a given load, of a cable net consisting of fixed cables, can be determined by bringing the equilibrium subspace containing the net positions calculable for all the arrangements of the forces H into "intersection" with the subspace of the cable positions compatible with the cable lengths belonging to the prescribed temperature. The intersection of these two spaces provides the cable position and forces H, i.e. the complete solution to the problem.

This procedure, simple in principle, can be illustrated in the case of a net consisting of, at most, two cables [23]. In the case of a net consisting of many cables, the calculation meets with difficulties, and, as for the computational aspect (i.e. for constructing the intersection of the subspaces in question), it is not possible to give a direct procedure that can be used in the general case.

In such cases, an obvious procedure is to determine the shape of the net corresponding to the changed load (and, of course, to the changed cable temperature) by iteration, starting from some basic position of the net (for example, from the erection state). However, in the case of cable nets, the iteration should be effected only very cautiously, and therefore it is desirable to analyze the change in the position and the cable forces of the net (i.e. "the change of state of the net") in more detail.

The space coordinates of the joints of a net with a spatially curved axis and consisting of cables fixed to a rigid edge, as well as the cable forces, are called the state characteristics of the net. The state characteristics, as elements of a multi-dimensional vector, determine one point of the so-called state space of the net. A relationship between the elements of each vector of the state space is created by the equilibrium and compatibility equations of the net. If the forces H and the coordinates of the joints of the cable net are considered to be the elements of the independent variable vector $\mathbf{r}$ of the state space, then the vector **v** (containing the cable temperatures and the concentrated external forces at the joints as elements) will form the dependent variable vector of the state space. The change of state of the net can be described by the explicitly expressible *non-linear* vector function $\mathbf{v}=\mathbf{v}(\mathbf{r})$. The local change of the state can be analyzed with the aid of the linear subspace $d\mathbf{v}=\mathbf{D}\cdot d\mathbf{r}$ that can be constructed at the point corresponding to the basic position. The elements of the matrix of the derived tensor $\mathbf{D}=\mathbf{D}(\mathbf{r})$, also required in the analysis, can easily be generated. It is clear from expressions (5.15) and (5.16), already discussed above, that the allocation $\mathbf{v}=\mathbf{v}(\mathbf{r})$ is unique:

$$\mathbf{t} = \mathbf{t}(\mathbf{H}, \mathbf{z}) \quad \text{and} \quad \mathbf{P} = \mathbf{P}(\mathbf{H}, \mathbf{z}),$$

where

$$\mathbf{r} = \begin{bmatrix} \mathbf{H} \\ \mathbf{z} \end{bmatrix}; \quad \mathbf{v} = \begin{bmatrix} \mathbf{t} \\ \mathbf{P} \end{bmatrix};$$

$$\mathbf{H} = \begin{bmatrix} H_{x1} \\ H_{x2} \\ \vdots \\ H_{xn} \\ H_{y1} \\ H_{y2} \\ \vdots \\ H_{ym} \end{bmatrix}; \quad \mathbf{z} = \begin{bmatrix} z_{11} \\ z_{12} \\ \vdots \\ z_{1n} \\ z_{21} \\ z_{22} \\ \vdots \\ z_{mn} \end{bmatrix}; \quad \mathbf{t} = \begin{bmatrix} t_{x1} \\ t_{x2} \\ \vdots \\ t_{xn} \\ t_{y1} \\ t_{y2} \\ \vdots \\ t_{ym} \end{bmatrix}; \quad \mathbf{P} = \begin{bmatrix} P_{11} \\ P_{12} \\ \vdots \\ P_{1n} \\ P_{21} \\ P_{22} \\ \vdots \\ P_{mn} \end{bmatrix}.$$

The elements of the matrix $\mathbf{D}(\mathbf{r})$ of the tangent-subspace belonging to point $(\mathbf{r}, \mathbf{v})$ of the state space of the cable net model presented in the foregoing can be generated on the basis of (5.15) and (5.16).
Let us introduce the following notations which will play an important role hereinafter:

$$\mathbf{D} = \begin{bmatrix} \mathbf{D}_{tH} & \mathbf{D}_{tz} \\ \mathbf{D}_{PH} & \mathbf{D}_{Pz} \end{bmatrix}$$

$$\mathbf{D}_{tH} = \begin{bmatrix} \left[\dfrac{\partial t_{xk}}{\partial H_{x\nu}}\right] & \left[\dfrac{\partial t_{xk}}{\partial H_{y\mu}}\right] \\ \left[\dfrac{\partial t_{yj}}{\partial H_{x\nu}}\right] & \left[\dfrac{t_{yj}}{H_{y\mu}}\right] \end{bmatrix}; \quad \mathbf{D}_{tz} = \begin{bmatrix} \left[\dfrac{\partial t_{xk}}{\partial z_{j,\nu}}\right] \\ \left[\dfrac{\partial t_{yj}}{\partial z_{\mu,k}}\right] \end{bmatrix};$$

$$\mathbf{D}_{PH} = \begin{bmatrix} \left[\dfrac{\partial P_{j,k}}{\partial H_{x\nu}}\right] & \left[\dfrac{\partial P_{j,k}}{\partial H_{y\mu}}\right] \end{bmatrix}; \quad \mathbf{D}_{Pz} = \left[\dfrac{\partial P_{j,k}}{\partial z_{\mu,\nu}}\right];$$

$(j, \mu = 1, 2, \ldots, m)$;
$(k, \nu = 1, 2, \ldots, n)$.

The relationship $\mathbf{dv}=\mathbf{D}\,\mathbf{dr}$ between the vector differentials $\mathbf{dv}$ and $\mathbf{dr}$, and its inversion can be expressed as follows:

$$\mathbf{dv}=\begin{bmatrix}\mathbf{dt}\\ \mathbf{dP}\end{bmatrix};\quad \mathbf{dr}=\begin{bmatrix}\mathbf{dH}\\ \mathbf{dz}\end{bmatrix}$$

$$\mathbf{dt}=\mathbf{D}_{tH}\,\mathbf{dH}+\mathbf{D}_{tz}\,\mathbf{dz}$$

$$\mathbf{dP}=\mathbf{D}_{PH}\,\mathbf{dH}+\mathbf{D}_{Pz}\,\mathbf{dz}$$

whence

$$\mathbf{dH}=\mathbf{D}_{tH}^{-1}(\mathbf{dt}-\mathbf{D}_{tz}\,\mathbf{dz}),$$

$$\mathbf{dz}=\mathbf{D}_{Pz}^{-1}(\mathbf{dP}-\mathbf{D}_{PH}\,\mathbf{dH}),$$

and finally

$$\mathbf{dH}=(\mathbf{D}_{tH}-\mathbf{D}_{tz}\mathbf{D}_{Pz}^{-1}\mathbf{D}_{PH})^{-1}(\mathbf{dt}-\mathbf{D}_{tz}\mathbf{D}_{Pz}^{-1}\mathbf{dP}).$$

If we use the fact that

$$\mathbf{H}=\mathbf{H}(\mathbf{t},\mathbf{z});\quad \mathbf{dH}=[\mathbf{D}_{Ht}\mathbf{D}_{Hz}]\begin{bmatrix}\mathbf{dt}\\ \mathbf{dz}\end{bmatrix},$$

$$\mathbf{z}=\mathbf{z}(\mathbf{H},\mathbf{P});\quad \mathbf{dz}=[\mathbf{D}_{zH}\mathbf{D}_{zP}]\begin{bmatrix}\mathbf{dH}\\ \mathbf{dP}\end{bmatrix},$$

then, with H=const., we can write

$$\mathbf{D}_{Pz}^{-1}=\mathbf{D}_{zP}\quad\text{and}\quad \mathbf{D}_{tH}^{-1}=\mathbf{D}_{Ht}^{\S}.$$

The elements of each column of the matrix $\mathbf{D}_{zP}$ can be obtained in such a manner that, with H=const. corresponding to the state under investigation, the values $z_{j,k}$ are calculated from the single load $P_{\mu,\nu}=1$ corresponding to the serial number of the column; and these then provide the elements of the $(\mu-1)n+\nu$th column of the matrix.

This procedure does not require the calculation of the elements of the derived tensor if, instead of vector differentials **dt**, **dP**, **dH** and **dz**, small finite increments are used and, from these, the increments **dH**, **dz** belonging to the prescribed **dt**, **dP** are determined. The appropriate computational procedure for the solution of the problem is as follows:

1. With the assumption of $\mathbf{dH}=0$, and on the basis of (5.15) and (5.16) the increments

$$\mathbf{dz}_P=\mathbf{D}_{zP}\,\mathbf{dP},$$

$$\mathbf{dt}_P=\mathbf{D}_{tz}\mathbf{D}_{zP}\,\mathbf{dP}$$

are obtained.

§ Incidentally, these relationships can also be written on the basis of the differentiation rules for functions of more than one variable.

2. With the assumption of $\mathbf{dP}=0$, the increments

$$\mathbf{dz}_H = -\mathbf{D}_{zP}\mathbf{D}_{PH}\,\mathbf{dH} = \mathbf{D}_{zH}\,\mathbf{dH}^{\dagger},$$

$$\mathbf{dt}_H = (\mathbf{D}_{tH}+\mathbf{D}_{tz}\,\mathbf{D}_{zH})\mathbf{dH}$$

are calculated.

This problem requires the solution of the system of equations

$$\frac{1}{a}(H_{xk}+\mathrm{d}H_{xk})(-z_{j-1,k}-\mathrm{d}z_{j-1,k}+2z_{j,k}+2\mathrm{d}z_{j,k}-z_{j+1,k}-\mathrm{d}z_{j+1,k})+$$

$$+\frac{1}{b}(H_{yj}+\mathrm{d}H_{yj})(-z_{j,k-1}-\mathrm{d}z_{j,k-1}+2z_{j,k}+2\mathrm{d}z_{j,k}-z_{j,k+1}-\mathrm{d}z_{j,k+1}) = P_{jk}$$

resulting from the equation system (5.16). If the products of the differentials are left out of consideration in this equation system and the zero valency corresponding to (5.16) is omitted from every equation, then the equation system serving for the calculation of the dz's corresponding to the prescribed dH's is obtained:

$$\frac{1}{a}H_{xk}(-\mathrm{d}z_{j-1,k}+2\mathrm{d}z_{j,k}-\mathrm{d}z_{j+1,k})+\frac{1}{b}H_{yj}(-\mathrm{d}z_{j,k-1}+2\mathrm{d}z_{j,k}-\mathrm{d}z_{j,k+1}) =$$

$$= \frac{1}{a}\mathrm{d}H_{xk}(-z_{j-1,k}+2z_{j,k}-z_{j+1,k})+\frac{1}{b}\mathrm{d}H_{yj}(-z_{j,k-1}+2z_{j,k}-z_{j,k+1}).$$

3. Since the prescribed temperature increment is

$$\mathbf{dt} = \mathbf{dt}_p+\mathbf{dt}_H,$$

the expression for the required $\mathbf{dH}$ is

$$\mathbf{dH} = (\mathbf{D}_{tH}+\mathbf{D}_{tz}\mathbf{D}_{zH})^{-1}(\mathbf{dt}-\mathbf{dt}_p).$$

This expression for $\mathbf{dH}$ agrees, of course, exactly with the expression

$$\mathbf{dH} = (\mathbf{D}_{tH}-\mathbf{D}_{tz}\mathbf{D}_{Pz}^{-1}\mathbf{D}_{PH})^{-1}(\mathbf{dt}-\mathbf{D}_{tz}\mathbf{D}_{Pz}^{-1}\,\mathbf{dP})$$

described in the foregoing.

The elements of the individual columns of the matrix

$$\mathbf{D}_{tH}+\mathbf{D}_{tz}\mathbf{D}_{zH},$$

which play an important role in the solution of the problem, can be determined numerically with approximate exactness in such a manner that, with $\mathbf{dP}=0$, a finitely small value — but differing from zero (e.g. the value of the original component H multiplied by 10^{-2}) — is chosen for each element of

† On the basis of the differentiation rules of implicitly given functions of more than one variable. Similarly: $-\mathbf{D}_{Ht}\mathbf{D}_{tz}=\mathbf{D}_{Hz}$.

the vector **dH**, and from this the *z* coordinates corresponding to the changed state of equilibrium are determined on the basis of (5.16); at the same time the temperature values **t** needed to ensure the compatibility of the changed state of equilibrium are determined on the basis of (5.15).

4. From a knowledge of **H**+**dH**, **z**+**dz** belonging to the given **P**+**dP** can be calculated on the basis of the equation system (5.16).

5. Finally, it is necessary to determine the error vector of the load, that is, — considering that the equilibrium equations were solved exactly for the prescribed load with the changed cable forces, and thus the error vector of the load is a zero vector —, the value of the temperature increments ensuring compatibility. The knowledge of the error vector is necessary for the next step in the investigation of the change of state (or for the decision concerning the termination of the calculation).

Example 5.4. The application of the procedure is illustrated by the following numerical example, which is performed on the cable net shown in Fig. 5.15:

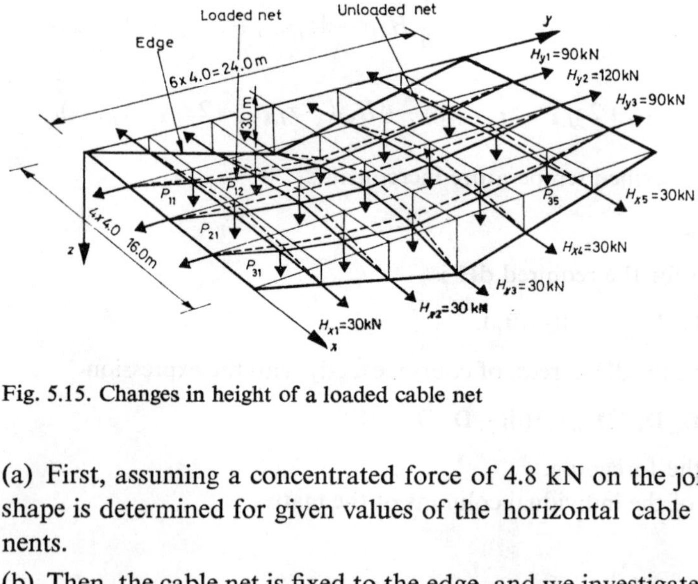

Fig. 5.15. Changes in height of a loaded cable net

(a) First, assuming a concentrated force of 4.8 kN on the joints, the cable shape is determined for given values of the horizontal cable force components.

(b) Then, the cable net is fixed to the edge, and we investigate in what cable position and under what cable forces the equilibrium position comes into being when the joint loads are increased by 2.0 kN.

The fact that the structure and the load possess double symmetry is utilized in the course of the calculation. Apart from this, the calculation is carried out on the basis of equation system (5.16).

The height data (including the edge) under the basic load is:

$$\begin{matrix} & 1.000000 & 2.000000 & 3.000000 & 2.000000 & 1.000000 & \\ 0 & 0.819481 & 1.409617 & 1.676912 & 1.409617 & 0.819481 & 0 \\ 0 & 0.687000 & 1.147759 & 1.317590 & 1.147759 & 0.687000 & 0 \\ 0 & 0.819481 & 1.409617 & 1.676912 & 1.409617 & 0.819481 & 0 \\ & 1.000000 & 2.000000 & 3.000000 & 2.000000 & 1.000000 & \end{matrix}.$$

Let us fix the net on the edge in the above basic position, and let us then increase the joint loads.

Data:

$E = 16\ \text{MN/cm}^2$

$A_{xk} = 1.2\ \text{cm}^2 \quad (k = 1, 2, \ldots, 5)$

$A_{y1} = 3.5\ \text{cm}^2$

$A_{y2} = 4.5\ \text{cm}^2$

$A_{y3} = 3.5\ \text{cm}^2$

$\alpha_t = 10^{-5}/1°\text{C}$

Step 1:

unchanged forces H
changed load: $\mathbf{P}+\Delta\mathbf{P}$
changed $\mathbf{z}$: $\mathbf{z}+\Delta\mathbf{z}$

to be calculated: $\Delta\mathbf{t}_p$ ensuring the compatibility of the system of internal forces and the displacement of the net.

$$\Delta\mathbf{z}_P = \begin{bmatrix} 0.1156321 & 0.1752275 & 0.1936571 \\ 0.1327074 & 0.2072856 & 0.2312263 \end{bmatrix}.$$

The temperature increments calculated on the basis of expression (5.15), under which the compatibility condition is satisfied, are:

$$\Delta\mathbf{t}^*_{xP} = [-72.187 \quad -345.023 \quad -793.701]\ °\text{C},$$

$$\Delta\mathbf{t}^*_{yP} = [274.700 \quad 266.044]\ °\text{C}.$$

Step 2:

variation of the cable forces (with increment $\Delta H = 0.01H$) and the determination of the $\Delta \mathbf{z}$ vectors coming into being under the effect of the former:

$\Delta H_{x1} = 0.3$

$$\Delta \mathbf{z} = \begin{bmatrix} 0.585967 & 0.553411 & 0.534900 \\ 1.426267 & 1.064667 & 0.958744 \end{bmatrix} \times 0.3 \times 10^{-3},$$

$\Delta H_{x2} = 0.3$

$$\Delta \mathbf{z} = \begin{bmatrix} 1.966033 & 4.321489 & 3.932067 \\ 2.763856 & 5.926678 & 5.527711 \end{bmatrix} \times 0.3 \times 10^{-3},$$

$\Delta H_{x3} = 0.3$

$$\Delta \mathbf{z} = \begin{bmatrix} 2.043067 & 4.613178 & 8.511822 \\ 2.504956 & 5.240733 & 8.290389 \end{bmatrix} \times 0.3 \times 10^{-3},$$

$\Delta H_{y1} = 0.9$

$$\Delta \mathbf{z} = \begin{bmatrix} -3.393367 & -5.763444 & -6.894844 \\ -2.211778 & -3.832778 & -4.445189 \end{bmatrix} \times 0.9 \times 10^{-3},$$

$\Delta H_{y2} = 1.2$

$$\Delta \mathbf{z} = \begin{bmatrix} -0.916378 & -1.554000 & -1.784556 \\ -2.669078 & -4.329178 & -4.952444 \end{bmatrix} \times 1.2 \times 10^{-3}.$$

Step 3:

determination of the matrix of compatible temperature increments

$$\mathbf{T}_H = \mathbf{D}_{tH} + \mathbf{D}_{tz}\mathbf{D}_{zH}$$

belonging to the changed cable forces, corresponding to the position depending on the ΔH's, on the basis of (5.15):

$$\mathbf{T}_H = \begin{bmatrix} -5.8814 & -1.4371 & -1.3417 & 1.4229 & 1.2406 \\ -1.4208 & -14.4009 & -8.9084 & 8.9369 & 5.0730 \\ -2.5384 & -17.0480 & -38.3611 & 24.3281 & 10.2512 \\ 0.9289 & 5.9014 & 8.6228 & -10.9376 & -2.4217 \\ 1.6325 & 6.7504 & 7.1836 & -4.8660 & -6.9496 \end{bmatrix}.$$

Since at present $\Delta \mathbf{t}=0$, we have

$$\Delta \mathbf{t} - \Delta \mathbf{t}_P = -\Delta \mathbf{t}_P = -\begin{bmatrix} \Delta \mathbf{t}_{xP} \\ \Delta \mathbf{t}_{yP} \end{bmatrix},$$

and

$$\Delta \mathbf{H} = \begin{bmatrix} \Delta \mathbf{H}_x \\ \Delta \mathbf{H}_y \end{bmatrix} = \mathbf{T}_H^{-1}(\Delta \mathbf{t} - \Delta \mathbf{t}_P),$$

where

$$\Delta \mathbf{H}_x^* = [-2.6580 \quad -5.9935 \quad -3.6917] \text{ kN},$$

$$\Delta \mathbf{H}_y^* = [14.8423 \quad 17.6277] \text{ kN}.$$

Step 4:
calculation of the heights $z+\Delta z$ corresponding to the joint effect of the forces H (increased by ΔH) and the load (increased by ΔP);

(a) If the Δz_p-values (calculated in step 1) and the product of the z's (calculated in step 2) which formed the H's (determined in step 3) are added to the z's corresponding to the basic load, then a very good approximation is obtained for the values $z+\Delta z$

$$\mathbf{z}+\Delta\mathbf{z}=\begin{bmatrix}0.847711 & 1.427487 & 1.680364\\ 0.710226 & 1.164146 & 1.329255\end{bmatrix},$$

where, e.g., the first element is

$$z_{1,1}+\Delta z_{1,1}=$$

$$=0.819481+0.115632-(2.6580\times0.585967+5.9935\times1.966033+3.6917\times$$

$$\times2.043067+14.8423\times3.393367+17.6277\times0.916378)=0.847711.$$

The $z+\Delta z$ coordinates determined in this way will balance, by means of the components $H+\Delta H$, the joint loads differing from the prescribed 6.8 kN value.

The values of the balanced loads are:

$$\begin{bmatrix}6.921 & 6.786 & 6.697\\ 6.939 & 6.795 & 6.743\end{bmatrix}\text{kN}.$$

(b) If the equation system (5.16) is solved *a priori* with the increased values of the cable forces

$$\mathbf{H}_x^*+\Delta\mathbf{H}_x^*=[27.3420\quad 24.0065\quad 26.3083]\ \text{kN},$$

$$\mathbf{H}_y^*+\Delta\mathbf{H}_y^*=[104.8423\quad 137.6277]\ \text{kN}$$

for the prescribed load

$$\mathbf{P}+\Delta\mathbf{P}=\begin{bmatrix}6.8 & 6.8 & 6.8\\ 6.8 & 6.8 & 6.8\end{bmatrix}\text{kN}$$

then, a more exact value of the $z+\Delta z$ coordinates is obtained:

$$\mathbf{z}+\Delta\mathbf{z}=\begin{bmatrix}0.845522 & 1.427217 & 1.678840\\ 0.707778 & 1.163254 & 1.328911\end{bmatrix}.$$

Step 5:
determination of the cable temperature increments ensuring compatibility (i.e. calculation of the error vector of the "load"):

$$\Delta\mathbf{t}_{xP}^*=[1.340\quad 1.016\quad 2.278]\ ^\circ\text{C}$$

$$\Delta\mathbf{t}_{yP}^*=[-2.043\quad -1.790]\ ^\circ\text{C}.$$

If this error vector is found to be greater than some permissible prescribed value, then the cable forces should be varied again; and, in accordance with this, the changed z's should be calculated (in the manner already described in the previous steps).

5.3.2.2. Net Fixed to an Elastic Edge

The computation of the model outlined for the investigation of the cable net is not unduly complicated by taking the elasticity of the edge also into account. Namely, according to our assumption, the horizontal joint displacements need not be taken into consideration in the equilibrium equations of the cable joints; hence, only the expressions of compatibility are affected by the elastic deformation of the edge.

We take the basic position of the edge to be the shape adopted by the latter under the effect of the cable forces arising in the basic (initial) position of the net (e.g. under dead-weight load). If this basic edge shape is regarded as given, then it is possible to determine the edge deformation due to cable forces which are equal and opposite to those in the actual system, i.e. we can determine the edge shape corresponding to the unloaded net. Thus, if, for example, under the basic load the cable edge has a rectangular ground plan according to Fig. 5.16a, then the edge of the unloaded net will assume the shape shown in Fig. 5.16b. In other words, the edge has to be designed according to the shape indicated in Fig. 5.16b in order that it should assume the shape shown in Fig. 5.16a under the basic load of the net.

The compatibility conditions become modified only in that, in the procedure of calculation discussed in the previous section (Step 2), it is necessary to determine, for the case $\Delta\mathbf{P}=0$, what edge movements occur during the variation of each force $\mathbf{H}$ (Fig. 5.16c), and the terms referring to the cable sections along the edge must be modified by these in (5.15); thus, for example, instead of expression

$$a\sum_{j=1}^{m+1}\frac{1}{\cos\alpha_{j,k}}$$

the following is to be written:

$$a_{1,k}\frac{1}{\cos\alpha_{1,k}}+a\sum_{j=2}^{m}\frac{1}{\cos\alpha_{j,k}}+a_{m+1,k}\frac{1}{\cos\alpha_{m+1,k}},$$

where $a_{1,k}$ and a_{m+1k} denote the lengths of the projections on the plane (x, y) of the sections of the kth cable along the edges $x=\text{const.}$

Thus, in the course of the calculation, the compatible temperatures which also take into consideration the edge displacements will be included in the matrix $\mathbf{T}_H$.

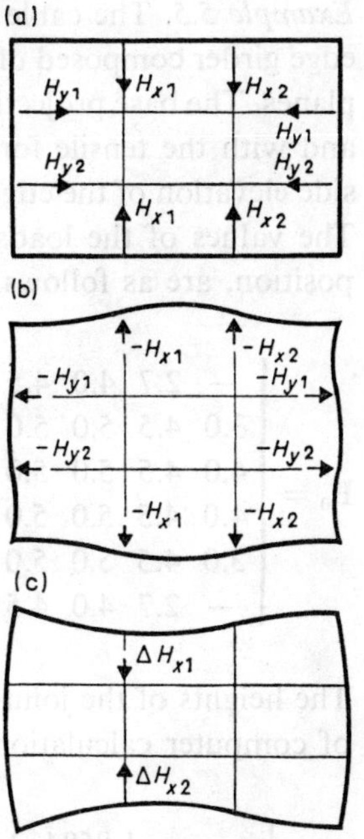

Fig. 5.16. Displacements of an elastic edge due to cable forces

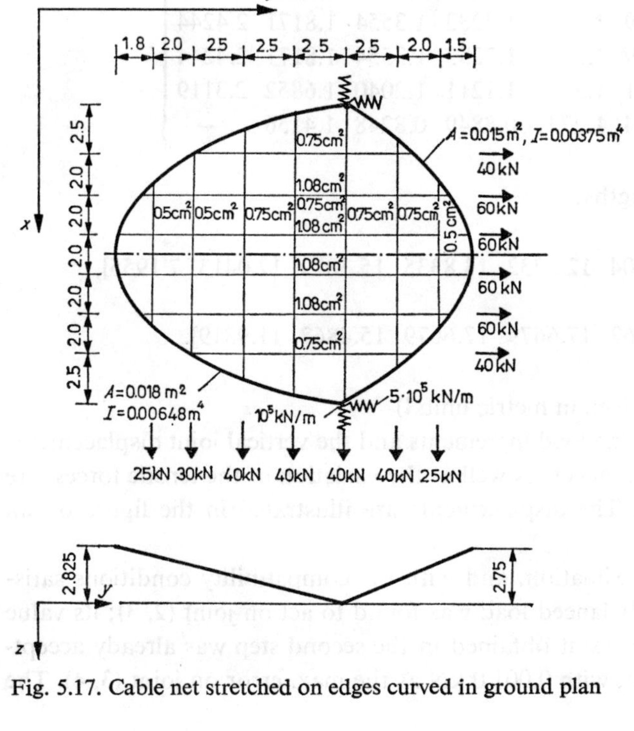

Fig. 5.17. Cable net stretched on edges curved in ground plan

Example 5.5. The cable net consists of 7×6 cables and is stretched on an edge girder composed of two parabolic arcs lying in two intersecting oblique planes. The base projection of the structure, with the geometrical dimensions and with the tensile forces prescribed in the basic position, as well as the side elevation of the edge girder, can be seen in Fig. 5.17.
The values of the loads, reduced to the joints corresponding to the basic position, are as follows (in kN):

$$\mathbf{P}_0 = \begin{bmatrix} - & 2.7 & 4.0 & 4.5 & 4.8 & 3.8 & - \\ 3.0 & 4.5 & 5.0 & 5.0 & 5.0 & 4.5 & 2.6 \\ 4.0 & 4.5 & 5.0 & 5.0 & 5.0 & 4.5 & 3.4 \\ 4.0 & 4.5 & 5.0 & 5.0 & 5.0 & 4.5 & 3.4 \\ 3.0 & 4.5 & 5.0 & 5.0 & 5.0 & 4.5 & 2.6 \\ - & 2.7 & 4.0 & 4.5 & 4.8 & 3.8 & - \end{bmatrix}.$$

The heights of the joints above the reference plane, according to the results of computer calculations, are:

$$\mathbf{z}_0 = \begin{bmatrix} - & 1.8584 & 1.2731 & 0.8849 & 0.8248 & 1.4156 & - \\ 2.3403 & 1.8251 & 1.3539 & 1.1211 & 1.2040 & 1.6852 & 2.3119 \\ 2.2928 & 1.8149 & 1.3987 & 1.2283 & 1.3554 & 1.8171 & 2.4244 \\ 2.2928 & 1.8149 & 1.3987 & 1.2283 & 1.3554 & 1.8171 & 2.4244 \\ 2.3403 & 1.8251 & 1.3539 & 1.1211 & 1.2040 & 1.6852 & 2.3119 \\ - & 1.8584 & 1.2731 & 0.8849 & 0.8248 & 1.4156 & - \end{bmatrix}.$$

The cables have lengths:

$$\mathbf{l}_x = [\ 7.5845 \quad 10.7704 \quad 12.3732 \quad 13.8335 \quad 15.3079 \quad 12.0413 \quad 7.1955],$$

$$\mathbf{l}_y = [11.9319 \quad 15.4862 \quad 17.6679 \quad 17.6679 \quad 15.4862 \quad 11.9319].$$

(The above data is given in metric units!)
The arrangement of the load increments and the vertical joint displacements occurring under their effect, as well as the variation of the tensile forces, are shown in Fig. 5.18. The displacements are illustrated in the figure on an enlarged scale.
After the first approximation, and with the compatibility conditions satisfied, the greatest unbalanced load was found to act on joint (2, 3); its value was 0.2106 kN. The result obtained in the second step was already acceptable as a final result, with 0.001 0 kN as the max. error on joint (3, 4). The

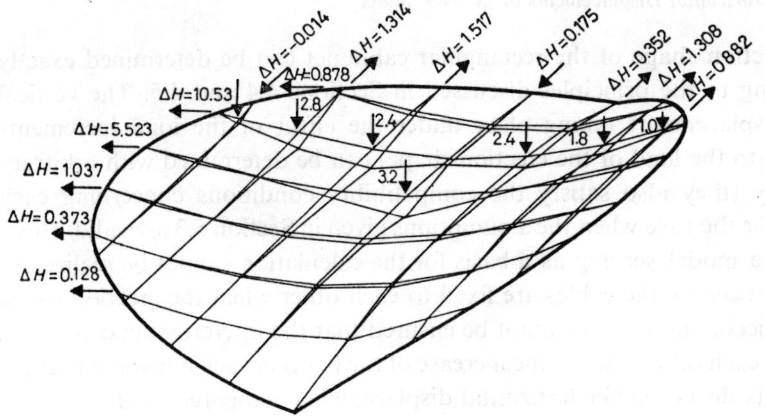

Fig. 5.18. Nodal displacements of a cable net due to loading

height positions of the joints were:

$$\mathbf{z}=\begin{bmatrix} - & 1.8565 & 1.2674 & 0.8833 & 0.8226 & 1.4140 & - \\ 2.3329 & 1.8026 & 1.3276 & 1.1156 & 1.1902 & 1.6751 & 2.3090 \\ 2.2943 & 1.8125 & 1.3867 & 1.1930 & 1.3508 & 1.8214 & 2.4294 \\ 2.2934 & 1.8142 & 1.3937 & 1.2193 & 1.3532 & 1.8181 & 2.4257 \\ 2.3405 & 1.8250 & 1.3522 & 1.1188 & 1.2027 & 1.6844 & 2.3118 \\ - & 1.8586 & 1.2731 & 0.8844 & 0.8240 & 1.4151 & - \end{bmatrix}.$$

Thus, the vertical displacements, from $\mathbf{z}=\mathbf{z}_0-\mathbf{z}$, are:

$$\Delta\mathbf{z}=\begin{bmatrix} - & 0.0019 & 0.0057 & 0.0016 & 0.0022 & 0.0016 & - \\ 0.0074 & 0.0225 & 0.0263 & 0.0055 & 0.0138 & 0.0101 & 0.0029 \\ -0.0015 & 0.0024 & 0.0120 & 0.0353 & 0.0046 & -0.0043 & -0.0050 \\ -0.0006 & 0.0007 & 0.0050 & 0.0090 & 0.0022 & -0.0010 & -0.0013 \\ -0.0002 & 0.0001 & 0.0017 & 0.0023 & 0.0013 & 0.008 & 0.0001 \\ - & -0.0002 & 0.0000 & 0.0005 & 0.0008 & 0.0005 & - \end{bmatrix}.$$

Because of the rather stiff supports and the great rigidity of the edge girder, the horizontal edge movement was only of the order of a few tenths of a millimetre; thus, its influence on the system of internal forces of the net was negligible. The coordinates of the edge girder belonging to the erection position differed from the coordinates of the basic position by a maximum of 2.0 mm.

5.3.2.3. Horizontal Displacements of the Net Joints

The erection shape of the rectangular cable net can be determined exactly according to the principles discussed in Sections 4.4 and 4.5. The vertical joint displacements taking place under the effect of the load increments (related to the load of the erection shape) can be determined with arbitrary accuracy (they also satisfy the compatibility conditions concerning each cable) for the case when the assumptions given in Section 5.3 are valid. However, the model serving as a basis for the calculation cannot be realized in practice because the cables are fixed to each other when the erection shape is produced, and thus it cannot be ensured that the intercrossing cables will slide on each other (during the increase of the load) in such a manner that the net joints do not suffer horizontal displacements compared to the erection position.

When the cables have been fixed to each other in the erection position, then, simultaneously with the vertical displacements, horizontal (x- and y-directional) displacements also occur at the joints; the approximate values of these can be calculated by a relatively simple procedure. The following assumptions are made in the course of the calculation:

1. The net joints are loaded only by a vertical excess force, i.e. $\Delta P_{xj,k} = \Delta P_{yj,k} = 0$ (compared with the erection load),
2. the values of the vertical net joint displacements $\Delta z_{j,k}$ due to the effect of the excess load do not change when the horizontal net joint displacements occur,
3. the intersection of horizontal joint displacements $\Delta x_{j,k}$ and $\Delta y_{j,k}$ can be neglected. In the course of the calculation, the erection shape determined according to Sections 4.4 and 4.5 is taken as a starting point; for the sake of simplicity, let us denote the lengths of the cable sections corresponding to this shape by l, in such a manner that

$$l_{xj,k} = \sqrt{a^2 + (z_{j,k} - z_{j-1,k})^2},$$

and

$$l_{yj,k} = \sqrt{b^2 + (z_{j,k} - z_{j,k-1})^2}.$$

The lengths of the cable sections, compared to the erection lengths, are altered under the effect of the temperature change (t_{xk}, and t_{yj}, respectively; $j = 1, 2, \ldots, m$; $k = 1, 2, \ldots, n$) prescribed for the individual cables and also under the cable force increments.

The joint displacements $(z_{j,k} + \Delta z_{j,k})$ and the horizontal cable force components $(H_{xk} + \Delta H_{xk},\ H_{yj} + \Delta H_{yj})$ of the state arising under the effect of the excess load ($\Delta P_{zj,k}$) determined according to the procedure described in Sections 5.3.2.1. and 5.3.2.2. (there is no horizontal displacement at the joints), are known. The cable lengths corresponding to this state can be calculated by means of expressions

$$\frac{a}{\cos\alpha_{j,k}}, \quad \frac{b}{\cos\beta_{j,k}},$$

where

$$\cos\alpha_{j,k} = \frac{a}{\sqrt{a^2+(z_{j,k}+\Delta z_{j,k}-z_{j-1,k}-\Delta z_{j-1,k})^2}},$$

$$\cos\beta_{j,k} = \frac{b}{\sqrt{b^2+(z_{j,k}+\Delta z_{j,k}-z_{j,k-1}-\Delta z_{j,k-1})^2}}.$$

Due to the temperature change t_{xk}, the length $l_{xj,k}$ of the section between joints $j-1, k$ and j, k of the kth x-directional cable (see Fig. 5.19a), corresponding to the erection state, changes to the value

$$l^{(1)}_{xj,k} = l_{xj,k}(1+t_{xk}\alpha_t).$$

Taking the horizontal displacement of the joints also into account, the geometrical length of the cable section will be

$$l^{(2)}_{xj,k} = \frac{a}{\cos\alpha_{j,k}} + (\Delta x_{j,k}-\Delta x_{j-1,k})\cos\alpha_{j,k}.$$

The difference between the lengths $l^{(2)}_{xj,k}$ and $l^{(1)}_{xj,k}$ originates from the elongation caused by the excess cable force,

$$\frac{\Delta H_{x,k}}{\cos\alpha_{j,k}}$$

that is

$$l^{(2)}_{xj,k} - l^{(1)}_{xj,k} = \frac{\Delta H_{xk}}{\cos\alpha_{j,k}}\,\frac{l_{xj,k}}{EA_{xk}}.$$

The quotient

$$\frac{\Delta H_{xk}}{EA_{xk}}$$

is identical at every section of the kth cable; thus

$$\frac{\cos\alpha_{j,k}}{l_{xj,k}}(l^{(2)}_{xj,k} - l^{(1)}_{xj,k})$$

must also be identical at every cable section. For the two cable sections con-

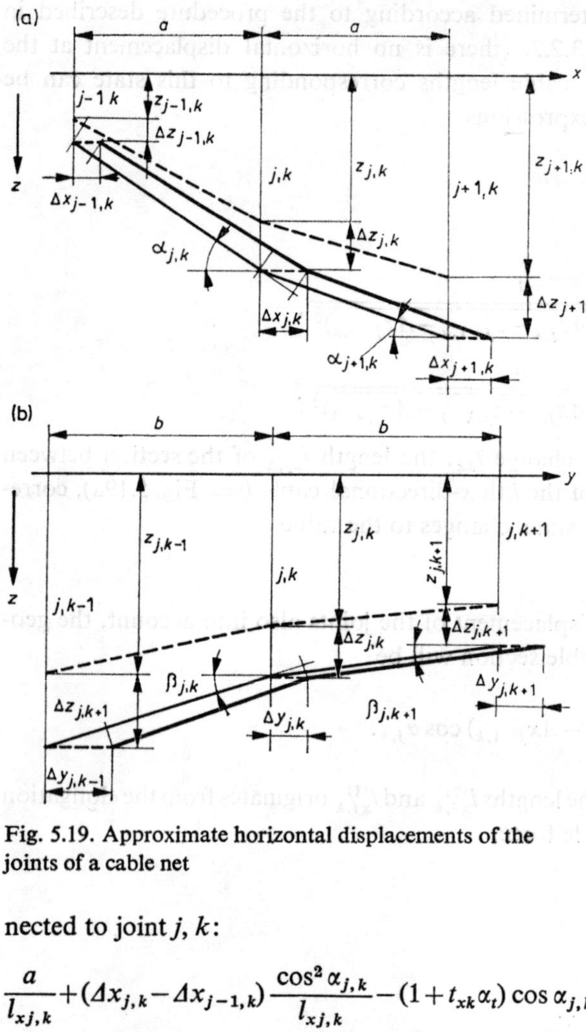

Fig. 5.19. Approximate horizontal displacements of the joints of a cable net

nected to joint j, k:

$$\frac{a}{l_{xj,k}}+(\Delta x_{j,k}-\Delta x_{j-1,k})\frac{\cos^2\alpha_{j,k}}{l_{xj,k}}-(1+t_{xk}\alpha_t)\cos\alpha_{j,k}=$$

$$=\frac{a}{l_{xj+1,k}}+(\Delta x_{j+1,k}-\Delta x_{j,k})\frac{\cos^2\alpha_{j+1,k}}{l_{xj+1,k}}-(1+t_{xk}\alpha_t)\cos\alpha_{j+1,k}$$

thus, the horizontal displacements $\Delta x_{j,k}$ ($j=1, 2, \ldots, m$) of the joints of the kth x-directional cable can be calculated from the equation system

$$-\Delta x_{j-1,k}\frac{\cos^2\alpha_{j,k}}{l_{xj,k}}+\Delta x_{j,k}\left(\frac{\cos^2\alpha_{j,k}}{l_{xj,k}}+\frac{\cos^2\alpha_{j+1,k}}{l_{xj+1,k}}\right)-$$

$$-\Delta x_{j+1,k}\frac{\cos^2\alpha_{j+1,k}}{l_{xj+1,k}}=a\left(-\frac{1}{l_{xj,k}}+\frac{1}{l_{xj+1,k}}\right)+$$

$$+(1+t_{xk}\alpha_t)(\cos\alpha_{j,k}-\cos\alpha_{j+1,k}). \tag{5.17}$$

A similar equation system serves for the calculation of the horizontal joint displacements Δy_j, $(k=1, 2, \ldots, n)$ of the jth y-directional cable:

$$-\Delta y_{j,k-1}\frac{\cos^2\beta_{j,k}}{l_{yj,k}}+\Delta y_{j,k}\left(\frac{\cos^2\beta_{j,k}}{l_{yj,k}}+\frac{\cos^2\beta_{j,k+1}}{l_{yj,k+1}}\right)-$$

$$-\Delta y_{j,k+1}\frac{\cos^2\beta_{j,k+1}}{l_{yj,k+1}}=b\left(-\frac{1}{l_{yj,k}}+\frac{1}{l_{yj,k+1}}\right)+$$

$$+(1+t_{yj}\alpha_t)(\cos\beta_{j,k}-\cos\beta_{j,k+1}). \tag{5.18}$$

The values of the horizontal cable displacements belonging to the rigid edge are known; hence the horizontal displacements of the joints of the individual x-directional (or y-directional) cables can be calculated with the aid of the system (5.17) or (5.18).

Example 5.6. Let us determine the horizontal joint displacements of the net stretched on a rigid edge discussed in Example 5.4, arising under the effect of the excess loads relative to the erection state. On the basis of the data calculated there:

j, k	$l_{xj,k}$	$\cos\alpha_{j,k}$	j, k	$l_{yj,k}$	$\cos\beta_{j,k}$
1,1	4.004 071	0.999 255	1,1	4.083 081	0.978 381
2,1	4.002 193	0.999 408	1,2	4.043 298	0.989 591
1,2	4.043 334	0.989 902	1,3	4.008 921	0.998 027
2,2	4.008 562	0.997 830	2,1	4.058 567	0.984 704
1,3	4.213 141	0.949 547	2,2	4.026 450	0.993 579
2,3	4.016 106	0.996 194	2,3	4.003 604	0.999 144

The temperature increment relative to the erection state is $t_{xk}=t_{yj}=0$ $(k=1, 2, \ldots, n;\ j=1, 2, \ldots, m)$. On account of the rigid edge and the double symmetry

$$\Delta x_{0,1}=\Delta x_{0,2}=\Delta x_{0,3}=\Delta x_{2,1}=\Delta x_{2,2}=\Delta x_{2,3}=0,$$

and

$$\Delta y_{1,0}=\Delta y_{2,0}=\Delta y_{1,3}=\Delta y_{2,3}=0,$$

so that the systems of equations for the determination of the horizontal displacements become simpler. For the first x-directional cable:

$$\left(\frac{\cos^2\alpha_{1,1}}{l_{x1,1}}+\frac{\cos^2\alpha_{2,1}}{l_{x2,1}}\right)\Delta x_{1,1}=a\left(-\frac{1}{l_{x1,1}}+\frac{1}{l_{x2,1}}\right)+\cos\alpha_{1,1}-\cos\alpha_{2,1},$$

from which

$$\Delta x_{1,1} = 0.006\,310,$$

and similarly

$$\Delta x_{1,2} = 0.001\,329,$$

$$\Delta x_{1,3} = -0.000\,154.$$

For the first y-directional cable:

$$\left(\frac{\cos^2\beta_{1,1}}{l_{y1,1}}+\frac{\cos^2\beta_{1,2}}{l_{y1,2}}\right)\Delta y_{1,1}-\frac{\cos^2\beta_{1,2}}{l_{y1,2}}\Delta y_{1,2} =$$

$$= b\left(-\frac{1}{l_{y1,1}}+\frac{1}{l_{y1,2}}\right)+\cos\beta_{1,1}-\cos\beta_{1,2},$$

$$-\frac{\cos^2\beta_{1,2}}{l_{y1,2}}\Delta y_{1,1}+\left(\frac{\cos^2\beta_{1,2}}{l_{y1,2}}+\frac{\cos^2\beta_{1,3}}{l_{y1,3}}\right)\Delta y_{1,2} =$$

$$= b\left(-\frac{1}{l_{y1,2}}+\frac{1}{l_{y1,3}}\right)+\cos\beta_{1,2}-\cos\beta_{1,3},$$

from where

$$\Delta y_{1,1} = -0.004\,330,$$

$$\Delta y_{1,2} = -0.002\,040,$$

and similarly:

$$\Delta y_{2,1} = -0.002\,653,$$

$$\Delta y_{2,2} = -0.001\,115.$$

By comparing the numerical results with the data obtained in Example 5.4, it can be seen that the greatest vertical joint displacement occurring under the effect of the excess load (relative to the erection state) took place at point 1, 1. Its value was $\Delta z_{1,1}=26.0$ mm. The greatest horizontal displacement also occurred at this joint; its absolute value was $\sqrt{\Delta x_{1,1}^2+\Delta y_{1,1}^2}=4.4$ mm.

5.3.3. Checking of the Results Obtained by the Method of Limited Accuracy

It was stressed during the introductory discussion of Section 5.3 that the model, serving as the basis of the procedure of limited accuracy, can be calculated with arbitrary accuracy, but that, when applied to the change of state of the real net, only approximate accuracy could be obtained.

As a result of the calculation, the values of the cable forces which arise under the effect of the vertical excess load (applied after the erection state is reached) and the values of the joint displacements were obtained. In the possession of this data, it is possible to calculate the joint loads which are in equilibrium in the new state, and the cable temperatures ensuring the compatibility of that state, with the aid of the procedure presented in Sections 5.1.2 and 5.2.2.

Example 5.7. With knowledge of the cable forces and joint coordinates of the net discussed in Examples 5.4 and 5.5, let us determine the magnitude of the loads necessary to ensure that equilibrium at the joints is satisfied. The cable forces are determined by the expressions

$$s_{xj,k} = \frac{H_{xk}+\Delta H_{xk}}{\cos\alpha_{j,k}}; \quad s_{yj,k} = \frac{H_{yj}+\Delta H_{yj}}{\cos\beta_{j,k}}.$$

Let us calculate the magnitude of the load required for the equilibrium of joint 1, 1.
The joint coordinates are:

j, k	$x_{j,k}$	$y_{j,k}$	$z_{j,k}$
0,1	0	4.0	1.0
1,1	4.000 631	3.995 670	0.845 522
2,1	8.0	3.997 347	0.707 778
1,0	4.0	0	0
1,1	4.000 631	3.995 670	0.845 522
1,2	4.001 329	7.997 960	1.427 217

The load ensuring equilibrium is:

$\mathbf{p}_{1,1} = \mathbf{A}\mathbf{s}$, where

$$\mathbf{A} = [\mathbf{e}_{0,1;1,1}\,\mathbf{e}_{2,1;1,1}\,\mathbf{e}_{1,0;1,1}\,\mathbf{e}_{1,2;1,1}] =$$

$$= \begin{bmatrix} 0.999255 & -0.999407 & 0.000154 & -0.000172 \\ -0.001082 & -0.000419 & 0.978336 & -0.989602 \\ -0.038585 & 0.034421 & 0.207025 & -0.143829 \end{bmatrix},$$

and

$$\mathbf{s} = \begin{bmatrix} s_{x1,1} \\ s_{x2,1} \\ s_{y1,1} \\ s_{y1,2} \end{bmatrix} = \begin{bmatrix} 27.36238 \\ 27.35820 \\ 107.15896 \\ 105.94508 \end{bmatrix} \text{kN}.$$

Thus,

$$\Delta \mathbf{p}_{1,1} = \begin{bmatrix} -0.0017 \\ -0.0471 \\ 6.8325 \end{bmatrix} \text{kN}.$$

The error in the load is:

$$\begin{bmatrix} 0.0017 \\ 0.0471 \\ -0.0325 \end{bmatrix} \text{kN}.$$

The error vectors corresponding to the other joints can also be calculated in a similar way.

Now, the cable temperatures ensuring the compatibility of the system must be determined for each cable section.

If, for example, the cable section between joints 1, 0 and 1, 1 of the net is investigated, and $l^{(0)}_{y1,1}$ and $l^{(1)}_{y1,1}$ denote the cable lengths belonging to the erection state and to the state corresponding to the excess load, respectively, and, moreover, if $s^{(0)}_{y1,1}$ and $s^{(1)}_{y1,1}$ denote the corresponding cable forces, then the compatibility of the state in this cable section is ensured by a temperature

$$t_{y1,1} = \frac{l^{(1)}_{y1,1} - l^{(0)}_{y1,1}\left(1 + \dfrac{s^{(1)}_{y1,1} - s^{(0)}_{y1,1}}{EA_{y1,1}}\right)}{l^{(0)}_{y1,1}\alpha_t}.$$

In our example, we have

$l^{(0)}_{y1,1} = 4.083\,081$ m $\quad l^{(1)}_{y1,1} = 4.084\,150$ m,

$s^{(0)}_{y1,1} = 91.869\,32$ kN, $\quad s^{(1)}_{y1,1} = 107.158\,96$ kN,

$E = 16$ MN/cm^2, $\quad A_{y1,1} = 3.5$ cm^2 $\quad \alpha_t = 10^{-5}/1$°C.

Thus

$t_{y1,1} = -1.09$ °C.

In other words, the length of the cable section under the effect of the excess cable force is greater by 0.0456 mm than the length calculated from the joint coordinates corresponding to the adopted state.

6. Supplementary Remarks

6.1. Buckling of the Edge Ring and Instability of the Cable Net

The edge ring may buckle under the effect of the compressive force arising in it. In the vertical plane, the buckling length is the distance between the supporting columns. In the horizontal plane (more formally, in the tangent plane of the cable surface) the cable net provides an elastic support to the ring. The extent of this elastic support depends on the sag-ratio (f_x/f_y) of the bidirectional cables, as will now be explained.

The force necessary for the deformation of the net in ground plan is measured by the internal force p arising between the cables, according to Section 2.4.3.2 [see formula (2.36)]. For a given Δl_x, this force p can differ in magnitude depending on the ratios f_x/f_y and l_x/l_y.

In the extreme case, when $p=0$, no force is required to produce the deformation of the cable net in ground plan. This is referred to as the "*geometrical instability*" of the cable net, and it occurs when the numerator of (2.36) becomes zero. By substituting the value of α from (2.21b), the condition of geometrical instability becomes:

$$\frac{f_x}{f_y} = \frac{l_x}{l_y} \frac{3 - \dfrac{l_x}{\sqrt{l_x l_y}}}{3 - \dfrac{l_y}{\sqrt{l_x l_y}}}. \tag{6.1}$$

Thus, in this case the edge ring buckles without being supported by the cable net at all. For other ratios f_x/f_y, the critical force for the ring elastically supported by the forces given by (2.37) must be determined (e.g. by the energy method); this value should then be compared to the (average) compressive force acting in the ring.

The geometrical instability can also be formulated in such a way that if the

edge ring has no flexural rigidity in the horizontal plane ($I_e=0$), then the structure offers no resistance against the deformation in the ground plan.
The phenomenon of the so-called "*statical instability*" of the cable net can also be illustrated by neglecting the flexural rigidity of the edge ring in the horizontal plane, in the following manner:
If the edge ring is regarded as infinitely flexible in the horizontal plane, then the ratio of the bidirectional cable forces must be such (for every loading case) that the edge ring is their funicular curve. This requirement makes it possible to determine the system of internal forces under dead weight (uniform loads) in a single step, the problem being statically determinate. Two equations can be written for the cable forces. On the one hand, relationship (2.3), expressing the funicular shape of the edge

$$\frac{n_x}{n_y}=\frac{l_x^2}{l_y^2}, \tag{6.2}$$

on the other, the vertical-projection equation, which can be written [on the basis of (1.1)] as follows

$$q=-8\left(n_x\frac{f_x}{l_x^2}-n_y\frac{f_y}{l_y^2}\right). \tag{6.3}$$

If one substitutes n_x from (6.2) into (6.3), the following expression is obtained for n_y:

$$n_y=-q\frac{l_y^2}{8(f_x-f_y)}. \tag{6.4}$$

From this it is immediately apparent that, in the special case when $f_x=f_y$ (independently of the ratio l_x/l_y), infinitely large cable forces arise, that is, the equilibrium conditions cannot be fulfilled by finite cable forces. Thus, the structure having a soft edge ring cannot be in equilibrium under a uniform load. For pretensioning, however, it can: according to formula (2.40), the pretensioning cable forces satisfy requirement (6.2). On the other hand, it is also true that the cable forces due to pretensioning satisfy condition (6.2) only in the case $f_x=f_y$ [cf. expression (2.40)]. Thus, a paradoxical situation arises: the cable net having a soft edge ring can neither support the pretensioning (if $f_x\neq f_y$), nor the uniform load (if $f_x=f_y$).
In the case of an edge ring of finite bending rigidity, the statical instability described above manifests itself in such a manner that in the case when $f_x=f_y$, very large edge moments arise under uniform load, whereas no bending moments arise from pretensioning.
The geometrical and static instabilities coincide if $l_x=l_y$, but in the case when $l_x\neq l_y$, they do not: e.g. for $l_x/l_y=0.5$, formula (6.1) gives $f_x=0.724f_y$.

6.2. The Optimum Shape of the Cable Net

On the basis of what has been said so far, the following question arises: what shape should be chosen for the cable net, so that the internal forces should be minimum?
We do not discuss the general problem, but we only point out a few tendencies. The results of the investigation are presented for the structure shown in Figure 3.1 on the basis of the calculations of Kollár and Köröndi [9] and Köröndi [9a]. Primarily, the effect of the sag-ratio of the pretensioning and supporting cables as well as the effect of the rigidity of the edge ring on the internal forces are considered. Our calculations were performed by the approximate method presented in Chapter 2, always taking the original geometry as a basis.

6.2.1. The Effect of the Sag-Ratio (f_x/f_y) of the Middle Pretensioning and Supporting Cables on the System of Internal Forces

In the following calculations, the sag of the supporting cable is kept at the constant value of $f_y = l_y/10 = 6.5$ m. The sag f_x of the pretensioning cable was varied between 2.17 and 6.5 m, thus the value of f_x/f_y ranged from 0.333 to 1.000. The rigidities of the edge and the cables are larger than in the previous numerical example, and are kept constant during our investigations:

$I_p = 1.35\ \text{m}^4 \qquad (EF_1)_x = 38\,200\ \text{kN/m}$

$E_p = 27.5\ \text{GPa} \quad (EF_1)_y = 46\,200\ \text{kN/m}.$

In every case the pretensioning force was chosen to be of such magnitude that the maximum "cable pressure" (in the case of dead weight+uniform wind suction) should be just compensated (exclusion of slackening: "minimum" pretensioning).
First of all, one should obtain the value of the ratio f_x/f_y for which the edge bending moments will be minimum. Namely, from the point of view of economy, the dimensions of the ring are more important than those of the cable cross-sections.
In Fig. 6.1 the variations of the moments arising at the bottom points of the edge arch are shown separately for the most unfavourable load types and load combinations as functions of the ratio f_x/f_y. It also follows from the figure that if the sag of the pretensioning cable is smaller than that of the supporting cable, only a moment with positive sign can be produced at the bottom point of the edge.

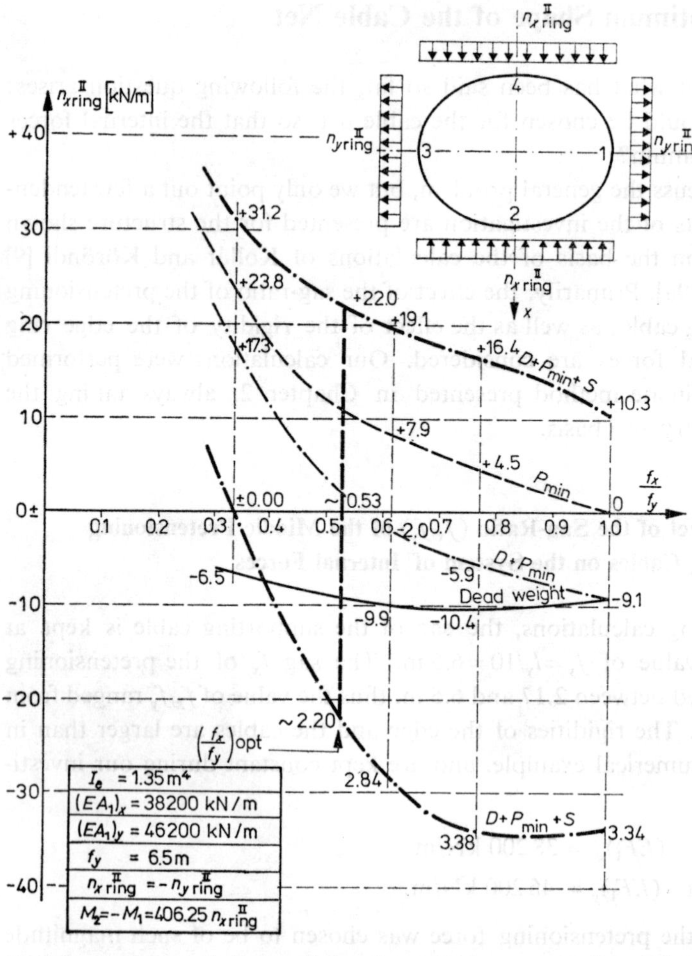

Fig. 6.1. Variation of the bending moment at the cross-section of the edge at the bottom point, as a function of the middle cable sag-ratio f_x/f_y, with unchanged cable and edge rigidities

The edge-bending moments M^-_{max} originating from the total snow load $(D+P+S)$ and M^+_{max} arising from uniform wind suction $(D+P+W_x)$ will be equal in absolute value in the case when $f_x/f_y \approx 0.53$. Hereinafter, this ratio will be regarded as "optimal"; in the case when such a ratio f_x/f_y is adopted, the minimum and optimum requirements of pretensioning are satisfied simultaneously; the resulting edge-bending moments will then be minimum in absolute value.

It appears from Figs 6.1 and 6.2, respectively, that the requirement $M^+_{max} = |M^-_{max}|$ can always be met in the case when $f_x/f_y \geqslant (f_x/f_y)_{opt}$ by the appli-

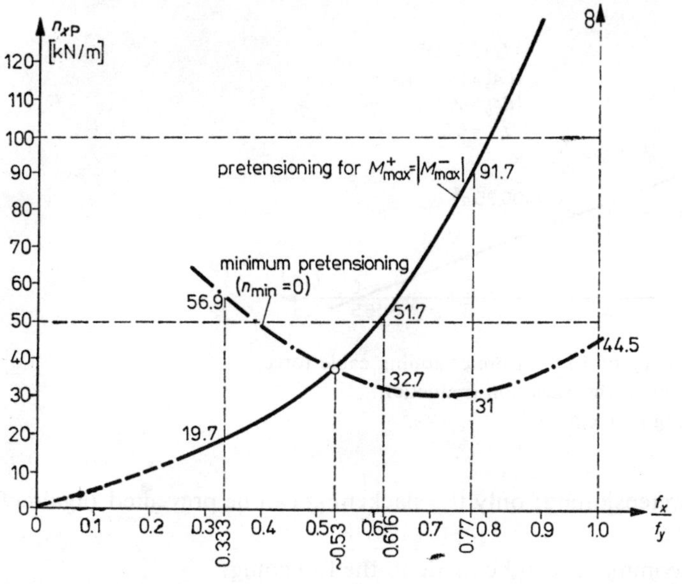

Fig. 6.2. Relationships between the pretensioning force to be applied in the pretensioning cables and the sag-ratio of the middle cables, for the cases of optimal and minimum pretensioning

cation of a pretensioning greater than that necessary for preventing slackening (upward flutter of cables), but never in the case when $f_x/f_y<(f_x/f_y)_{opt}$.
The relationship between the sag-ratios and the pretensioning cable forces n_{xP} necessary to satisfy the requirement $M^+_{max}=|M^-_{max}|$ and to prevent the slackening of the cables is indicated in Fig. 6.2.
If the sags of the supporting and pretensioning cables are of the same magnitude ($f_x/f_y=1$), then the bending of the edge cannot be influenced by pretensioning, i.e. the pretensioning is the funicular load of the net, causing pure compression in the edge, $M_P=0$ (cf. what was said in Section 6.1).
During design it should be remembered that when $f_x \approx f_y$, the equality oj the moment maxima (in absolute value) can be achieved only by huge pretensioning forces, which is uneconomical as regards the cables. From the view-point of the reduction of both the maximum cable forces and the edge-bending moments, it is desirable to adopt the "optimal" ratio f_x/f_y.
Figure 6.3 illustrates the phenomenon from another point of view: here the component of the 1 kN/m tensile cable force causing edge-bending, i.e. the magnitude of the edge-bending itself, is shown as a function of the ratio f_x/f_y. It can be seen that if $f_x/f_y \to 1$, then the moments cannot be influenced

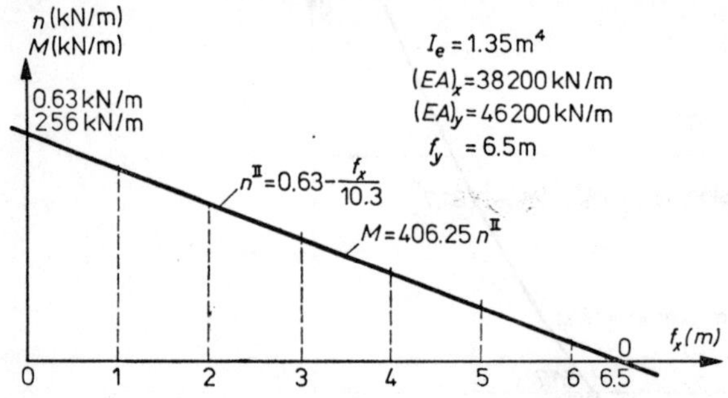

Fig. 6.3. The efficiency of 1 kN/m pretensioning cable force from the view-point of the variation (reduction) of the edge-bending moment

(reduced) by pretensioning; only the slackening can be prevented (again cf. Section 6.1).

The following comments can be made to the foregoing:

(a) In the course of the variation of the ratio f_x/f_y, not only the magnitude of the cable forces arising from the different loads can change to a significant extent, but their *sign* may also be altered. For example, in the case of a sag-ratio less than a certain f_x/f_y value, cable tension arises in both directions under uniform loads acting downwards, and "cable pressure" from loads acting upwards; on the other hand, at higher ratios f_x/f_y, "pressure" arises in the pretensioning cable for loads acting downwards. The same phenomenon can also be observed as I_e increases. In our numerical example, the "optimal" ratio f_x/f_y shown in the foregoing approximates the latter "limit" f_x/f_y quite well. Thus, in this case the pretensioning cables of the suspended roof do not take part in carrying the loads, or at least they take a very small portion of the load.

(b) In the course of our present investigation, the effect of the variation of the ratio f_x/f_y on the *shape factors* of the wind loads was not taken into consideration.

(c) The values of l_x, l_y, f_y were considered to be given. f_x may be also prescribed within certain limits; in this case, the optimum must be found within these limits.

In the case of the data adopted in the numerical example, it is no longer advisable to strive for the same magnitude of the maximum edge moments of opposite signs if $f_x/f_y \lesssim 0.75$, because of the large cable forces which must be applied in order to achieve them.

(d) In the course of the variation of the ratio f_x/f_y, different load combinations proved to be most unfavourable for the slackening of the pretensioning and supporting cables, respectively, i.e. the minimum pretensioning had to be determined from cable loadings originating from different load combinations. In the case of a relatively low sag-ratio $[f_x/f_y<(f_x/f_y)_{opt}]$, the minimum pretensioning is determined by the elimination of the "cable pressure" which arises in the supporting cable due to uniform wind suction. Inasmuch as the sag-ratio was higher than the above, we found that the pretensioning cables were prone to slackening, in particular under the y-directional wind load in the range of the sag-ratio $(f_x/f_y)_{opt}<f_x/f_y \lesssim 0.75$, and under the snow load when $f_x/f_y>0.75$.

6.2.2 Relationship between the Optimal f_x/f_y Ratio and the Relative Sag of the Supporting Cable (f_y/l_y)

Suspended roofs with different supporting cable sag are investigated in the case of identical edge ground plan ($l_x=50$ m, $l_y=65$ m) and identical cable and edge rigidities $[(EA_1)_x=38\,200$ kN/m, $(EA_1)_y=46\,200$ kN/m, $I_e=$ $=1.35$ m^4]. Thus, for differing values of f_y/l_y, it is possible to determine in each case the values $(f_x/f_y)_{opt}$ in the manner described in the previous section.

On the basis of numerical examples the following can be established:

(a) on roofs where the supporting cable sag conforms to $1/5 \geq f_y/l_y \geq 1/20$ the value of $(f_x/f_y)_{opt}$ is practically independent of the ratio (f_y/l_y);
(b) if the sag of the supporting cables is reduced, cable forces, edge moments and roof deformations increase.

In Fig. 6.4 the maximum moments arising in the case of the optimal ratio f_x/f_y are indicated in the range f_y/l_y investigated. Thus, the extent of the

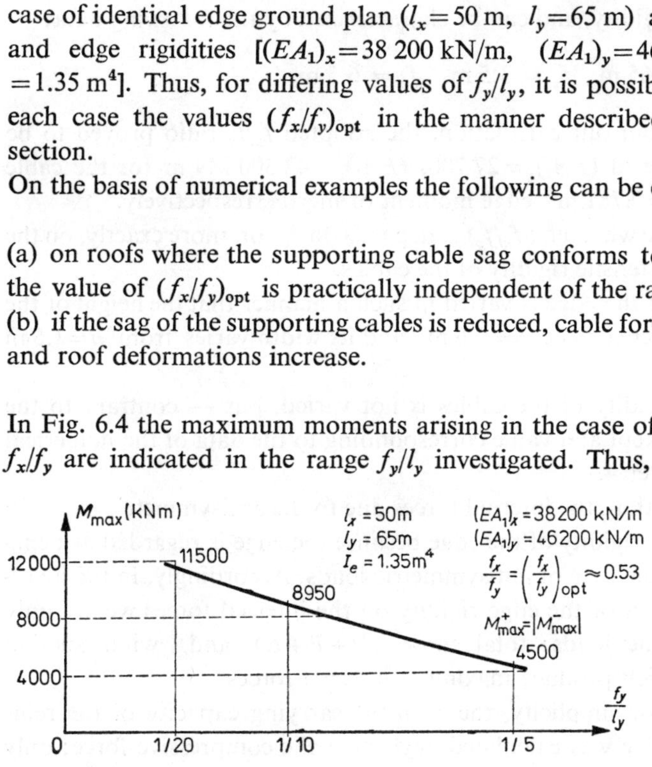

Fig. 6.4. The maximum edge-bending moment as a function of the supporting cable sag for the case of the optimal f_x/f_y ratio

minimum pretensioning and the "optimal" pretensioning, necessary for $M^+_{max}=|M^-_{max}|$, is identical for every point of the figure.

Notes:

(a) The effect of the antisymmetric loads — in the case of varying f_y/l_y — also varies. This has not been taken into consideration.
(b) When the ratio f_y/l_y is increased, the approximations that the cables are "flat" and that the edge is considered to be located in the horizontal plane become less accurate.

6.2.3. The Effect of Edge Rigidity on the System of Internal Forces

Our investigations are performed as follows:

(a) The original geometry of the cable net is kept constant during the variation of the edge rigidity, and can be characterized by the following data:

$l_x = 50$ m, $l_y = 65$ m, $f_x = 3.5$ m, $f_y = 6.5$ m.

As it appeared from our calculation, the adopted f_x/f_y ratio proved to be optimal in the case of $(EA_1)_x=27\,700$, $(EA_1)_y=42\,300$ kN/m for the cable rigidities and $I_e=0.781$ m^4 edge moment of inertia, respectively.
We shall determine whether $(f_x/f_y)_{opt}$ depends on I_e, or, more exactly, on the ratio of I_e and the tensile rigidity of the cables.
(b) The rigidity of the edge is varied in such a manner that the height of the edge ring always remains at $b=0.6$ m, and its width varies from $h=1.5$ m to 3 m.
(c) The tensile rigidity of the cables is not varied, but — contrary to the foregoing — it is kept at a value corresponding to the data of the numerical example of Chapter 4.
(d) In our calculation, the internal forces due to the antisymmetric loads do not depend on the rigidity of the edge because the edge is regarded as being infinitely rigid in the case of antisymmetric loads. Accordingly, in the investigation of the effect of the edge rigidity on the internal forces we are only concerned with the loads: total snow $(D+P+S)$ and wind suction $(D+P+W_x)$, which produce maximum internal forces.
(e) For the sake of simplicity, the moment carrying capacity of the reinforced concrete edge was calculated neglecting the compressive force; only tension reinforcement was used, and the concrete is B 200 (20 N/mm^2 cube strength).

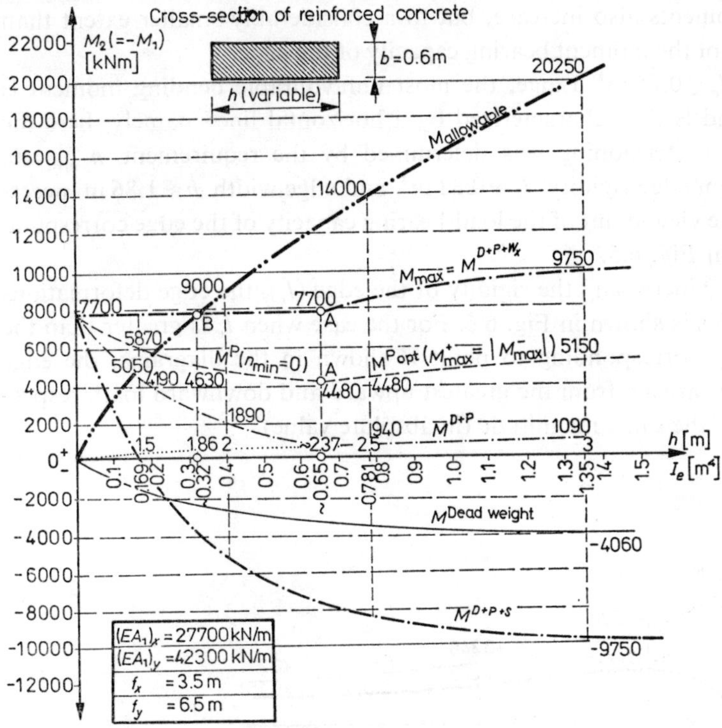

Fig. 6.5. Relationship between the bending moment at the bottom point of the edge and the moment of inertia of the edge for different loads, and for unchanged cable rigidities and roof geometry

In Fig. 6.5 the allowable bending moment of the reinforced concrete edge and the values of the edge-bending moment arising for the different loading cases are shown as functions of the moment of inertia and width of the edge, respectively.

In the case of I_e's smaller than the edge rigidity characterized by point A in the figure, the pretensioning which is optimum from the view-point of the edge moment $[M^{+}_{max}=|M^{-}_{max}|]$ was not sufficient for eliminating the cable pressure. In other words; the reduction of I_e increased the value of $(f_x/f_y)_{opt}$. Accordingly, the value of $(f_x/f_y)_{opt}$ depends on I_e as well: when I_e is increased for unchanged cable rigidities, as well as when the cable rigidities are reduced while I_e is kept constant, the value of $(f_x/f_y)_{opt}$ decreases; and it increases when I_e is decreased and the cable rigidities are increased. It is to be noted that in the practical range of dimensions, this decrease (or increase) is not considerable.

It can clearly be seen from the figure that when the edge rigidity is increased, the edge moments also increase, but to a considerably smaller extent than the increase of the moment bearing capacity of the edge.

In the $0<I_e\lesssim 0.65\ \mathrm{m}^4$ range, the most unfavourable bending moment is constant, and is thus characterized by a horizontal line; namely, here the intensity of pretensioning was determined by the requirement $n_{y\min}=0$. The minimum edge rigidity $I_e\approx 0.32\ \mathrm{m}^4$ and edge width $h\approx 1.86$ m necessary from the view-point of the load-bearing capacity of the edge correspond to point B in Fig. 6.5.

Of course, by increasing the rigidity of the edge (I_e), the edge deformations decrease. This is shown in Fig. 6.6. For the case when I_e is greater than the edge rigidity corresponding to point A shown in the diagrams, the edge deformations arising from the greatest upward and downward load, respectively, are of the same magnitude (in absolute value).

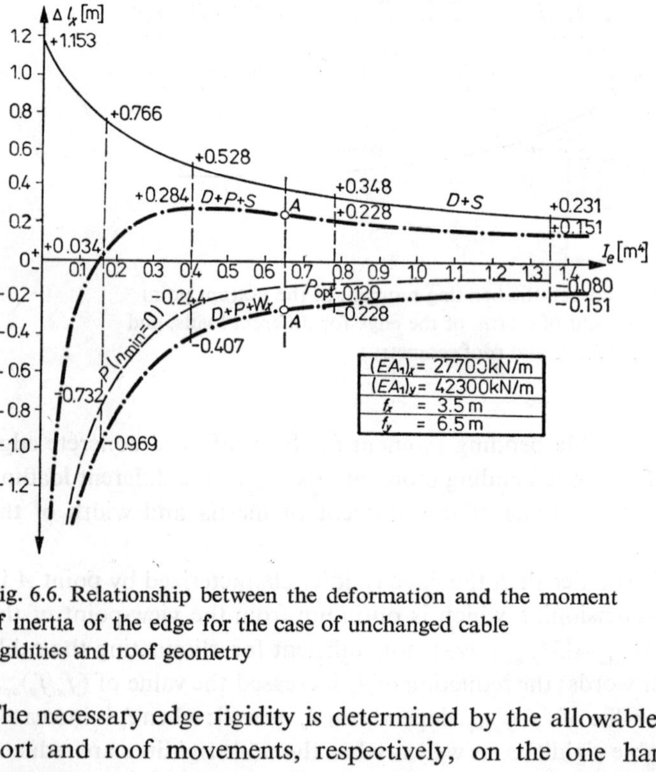

Fig. 6.6. Relationship between the deformation and the moment of inertia of the edge for the case of unchanged cable rigidities and roof geometry

The necessary edge rigidity is determined by the allowable maximum support and roof movements, respectively, on the one hand and by the moment- and compression-bearing requirements on the other.

On the basis of Fig. 6.6, the minimum necessary edge rigidity can be determined according to the permissible maximum edge deformation. Thus, this figure complements Fig. 6.5, which illustrated the load-bearing requirement.

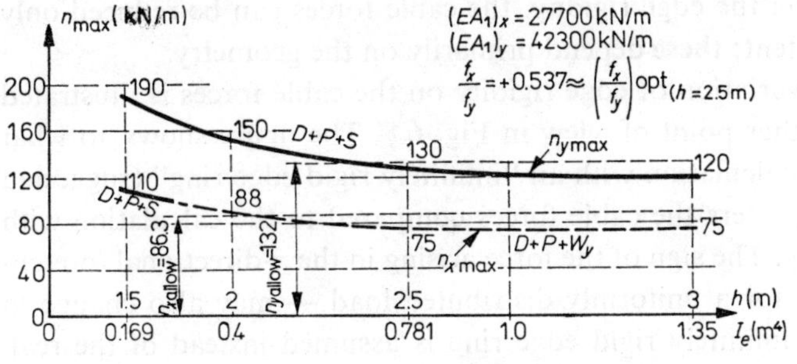

Fig. 6.7. Relatioship between the maximum cable forces and edge rigidity

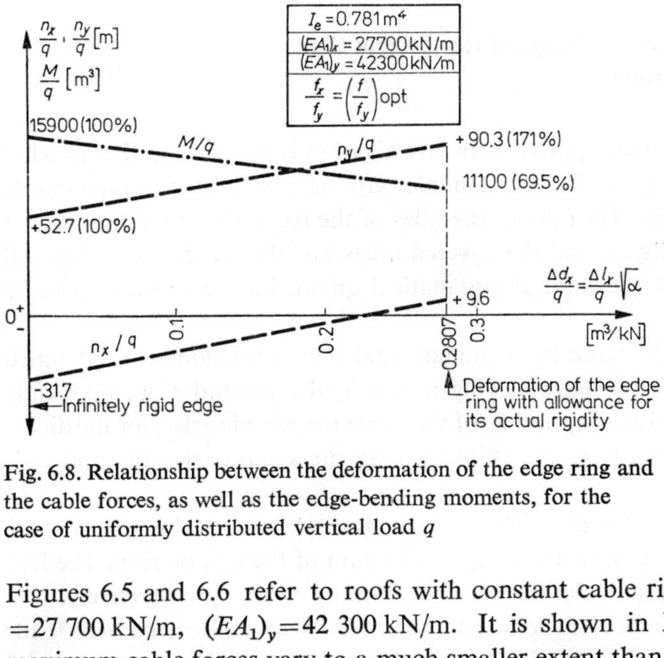

Fig. 6.8. Relationship between the deformation of the edge ring and the cable forces, as well as the edge-bending moments, for the case of uniformly distributed vertical load q

Figures 6.5 and 6.6 refer to roofs with constant cable rigidities $(EA_1)_x = 27\,700$ kN/m, $(EA_1)_y = 42\,300$ kN/m. It is shown in Fig. 6.7 that the maximum cable forces vary to a much smaller extent than the edge rigidity, and thus this approximation is acceptable. In other words: even if the necessary cable rigidity for the given problem is adopted, the principal values of the diagrams do not change significantly.

In the sections of the curves marked by a full line the load combination $D+P+S$ was the most unfavourable; in the section marked by a dash line, the load combination $D+P+W_y$ was least favourable for the cables.

According to Fig. 6.7 the cable rigidities chosen in the given numerical example are sufficient with the exception of the range $h \lesssim 2.3$ m; the load-bearing and deformation requirements, however, require just $h \gtrsim 2.3$ m for the edge widths.

Thus, by increasing the edge rigidity, the cable forces can be reduced only to a very small extent; these depend primarily on the geometry.
The effect of the variation of edge rigidity on the cable forces is illustrated according to another point of view in Fig. 6.8. The figure shows to what a large extent the calculation with an "infinitely rigid edge ring" (suggested by several papers) alters the cable forces compared to the calculation with an elastic edge ring. The sign of the force arising in the x-directional (pretensioning) cables — for a uniformly distributed load — may also change to its opposite if an infinitely rigid edge ring is assumed instead of the real, elastic one.

6.2.4. The Effect of the Shape of the Ground Plan on the Edge Moments

Finally, another brief, approximate investigation is made regarding to what extent the edge ring having a rectangular ground plan is more unfavourable than the elliptic one. The ratio of the sides of the rectangle equals the ratio of the axes of the ellipse, and the covered areas are also of the same size. All other geometrical, mechanical and statical quantities are considered to be identical.
The comparison is made by a fundamental approximation: it is assumed that in the case of an edge having a rectangular ground plan, uniformly distributed cable forces n_x and n_y of the same magnitude arise, for the different loading cases, as on the ellipse having the same basic area and axis length ratio.
The funicular lines for the different loads were drawn on the ground plan of the ellipse; these show the moment diagram of the elliptic ring. The half axis lengths of the funicular lines were marked by a_t and b_t, respectively.
The three most important load combinations were chosen, namely loads $D+P+S$, $D+P+W_x$ causing the maximum elliptic edge moments, and load $D+P$ which produces minimum edge moments.
In Figs 6.9–6.11 the rectangular edge was drawn, and the moments in the top and bottom points of the elliptic edge, as well as the field and corner moments of the rectangular edge are shown.
The maximum moments increased from +7980 kNm to +15 710 kNm for the load combination $D+P+S$ (Fig. 6.9) at the vertex, and from +7980 kNm to +9010 kNm for the load combination $D+P+W_x$ (Fig. 6.10) at the bottom point. The corner moments are not greater than the maximum field moment.
Another essential difference exhibited by the rectangular edge is that it

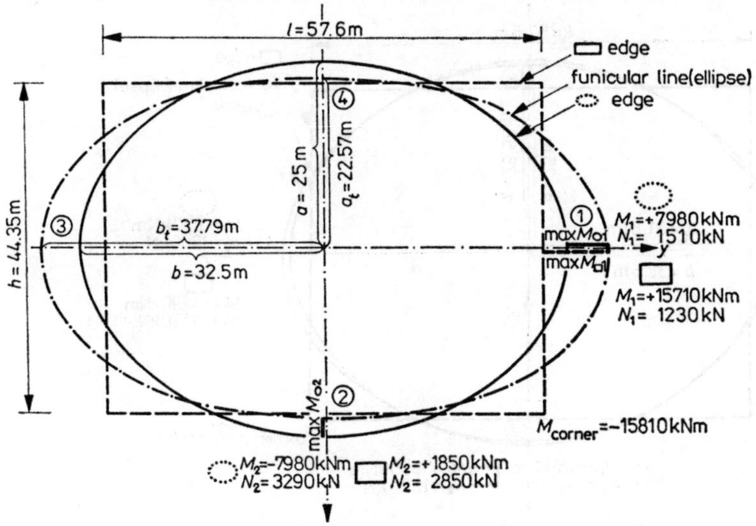

Fig. 6.9. Comparison of edge bending moments and compression for the case of edges with elliptic and rectangular ground plans, under $D+P+S$ load

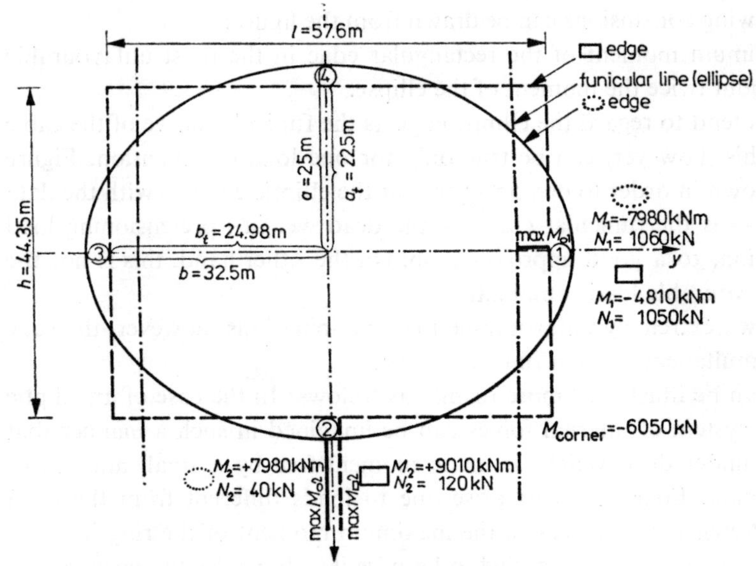

Fig. 6.10. Comparison of edge bending moments and compression for the case of edges with elliptic and rectangular ground plans, under $D+P+W_x$ load

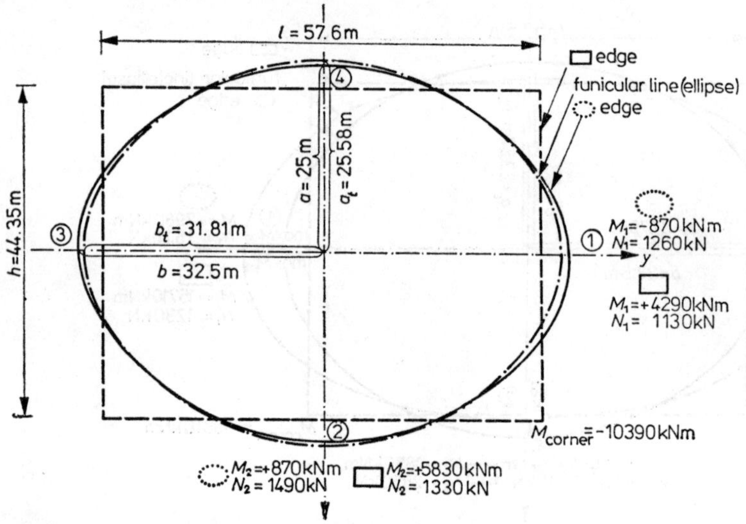

Fig. 6.11. Comparison of edge bending moments and compression for the case of edges with elliptic and rectangular ground plans, under $D+P$ load

suffers considerable torsion. This might be taken by the bending of the supporting columns.

The following conclusions can be drawn from the figure:

The maximum moment of the rectangular edge in the most unfavourable case is about twice the moment of the ellipse.

Designers tend to regard the elliptic edge as the funicular curve of the cable forces. This, however, can be true only for one load combination. Figure 6.11 is shown in order to demonstrate that the elliptic edge — with the data adopted — is the funicular curve of the dead weight+pretensioning load combination, to a good approximation. On the other hand, this is not the most unfavourable load combination.

For the two extremely unfavourable load combinations, however, the edge cannot simultaneously be a funicular curve.

All this can be illustrated more readily as follows: In the case of an elliptic edge, the system of internal forces can be imagined in such a manner that the edge under dead weight is made moment-free by a small amount of pretensioning. Edge moments arise due to loads different from the dead weight. Pretensioning increases the maximum moment of the ring if a sag-ratio less than $(f_x/f_y)_{opt}$ is applied, otherwise it only brings the positive and negative moment maxima nearer to each other.

In the case of a rectangular ground plan, the ring cannot be made moment-free for the dead weight by pretensioning. The moment arising from this

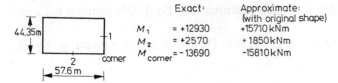

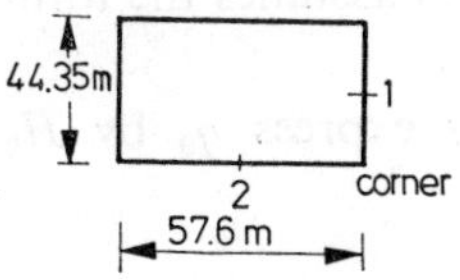

	Exact:	Approximate: (with original shape)
M_1	= +12930	+15710 kNm
M_2	= +2570	+ 1850 kNm
M_{corner}	= - 13690	-15810 kNm

Fig. 6.12. Comparison of edge moments for the suspended roof with rectangular ground plan, for the $D+P+S$ loading case

represents a surplus compared to the elliptic ring; this surplus, however, is not great. Though the moments arising from the two other effects (i.e. from the other loads and from pretensioning) will be greater than on the elliptic ring, they will not be their multiples. Thus, the approximately double moment magnitude obtained as a result of the investigation becomes understandable.

To check the approximate calculation, in Fig. 6.12 the approximate values of the edge moments of the structure having a rectangular ground plan are compared to the results obtained by the exact calculation, for the case of dead weight+pretensioning+total snow load. The numerical values show that the results of the approximate calculation made by taking the original shape as a basis differ from those of the exact one by about as much as in the case of the elliptic ground plan (Table 3.4). Thus, the conclusions drawn from the results of the above approximate calculation can be accepted as being correct.

6.3. The Local Flutter of the Cable Net

As was mentioned in Chapter 1, it would be desirable if the cable net were to be given such a shape that all parts of it should be — at least approximately — "uniformly rigid", i.e. that they should deform by the same amount under a given load. In this way, we anticipate that no local "flutter" will occur on the net due to local wind-load peaks, which, e.g., might lead to subsequent reinforcement, as in the Raleigh Arena.

Lacking more detailed investigations, the local rigidity of the cable net can be measured as follows:

Let us take one cable fixed at its two ends, as shown in Fig. 1.3 and let us consider the deflection w_k occurring under the effect of the load Δq; or, more exactly, the differential quotient of w_k with respect to Δq is to be the index number of rigidity.

Besides Eq. (1.3), the compatibility equation (1.29) is also necessary for the determination of w_k. Since the shape of the cable is a flat second-degree

parabola both before and after deformation, Eq. (1.29) assumes the form of (2.26).
Let us substitute ΔH from (2.26) into (1.3), and let us express q_0 by H_0 according to (1.1):

$$\left(H_0+\frac{16}{3}\frac{EA_f w_k}{l^2}+\frac{8}{3}\frac{EAw_k^2}{l^2}\right)(f+w_k) = H_0 f+\frac{\Delta q l^2}{8}. \tag{6.5}$$

Introducing the curvature of the cable according to (1.15), let us arrange the equation in accordance with the powers of w_k:

$$\frac{8EA}{3l^2}w_k^3+EAkw_k^2+\left(H_0+\frac{EAl^2}{12}k^2\right)w_k = \frac{\Delta q l^2}{8}. \tag{6.6}$$

Denoting now the differentiation with respect to Δq by $'$, let us differentiate (6.6) and obtain w_k':

$$w_k' = \frac{l^2/8}{\frac{8EA}{l^2}w_k^2+2EAkw_k+\left(H_0+\frac{EAl^2}{12}k^2\right)}, \tag{6.7}$$

which yields the expression

$$w_k'(0) = \frac{l^2/8}{H_0+\frac{EAl^2}{12}k^2} \tag{6.8}$$

for the initial rigidity at $w_k=0$.
The following conclusions can be drawn from Eq. (6.8):
In order that the rigidity of the cable should not change, a greater curvature is necessary in the case of a smaller cable force, and vice versa. If the cable force and the curvature become simultaneously zero, then the cable system does not display any initial rigidity against the loads acting on it, that is, it will certainly develop flutter.
On the basis of the foregoing, of the erection shapes discussed in Section 1.2.2.3 those can be considered to be "favourable" from the view-point of flutter where, on the one hand, both the cable force and the curvature are constant [hyperbolic paraboloid: Eq. (1.14)], and, on the other hand, those for which the curvature becomes zero at certain parts, (nevertheless here the cable forces are greater than in the portions of finite curvature [the surface characterized by Eq. (1.24)], but the translation surfaces consisting of hyperbolic cosine curves and circular arcs, respectively, in both directions, may also be classified in this group). Those surfaces having a curvature which

decreases in some parts without an increase in the cable force [e.g. the surfaces described by Eqs (1.22) and (1.23)] and even more so the ones for which the curvature and the cable force decrease simultaneously, may be classified as "unfavourable" from the point of view of flutter.

6.4. Vibrations of the Entire Cable Net

The natural circular frequencies ω of the cable net constructed above a rectangular ground plan were determined approximately by Schleyer [17] — taking an infinitely rigid edge as a basis — for the following three vibration forms:

(a) Symmetrical vibration form (with one half-wave in both directions):

$$\omega_{ss}^2 = \frac{\pi^2}{\mu}\left[\frac{H_x}{lx^2} + \frac{H_y}{ly^2}\right]. \tag{6.9}$$

Here and hereinafter, μ denotes the mass of the cable net per unit area.
(b) Vibration form symmetrical in one (y) direction, antisymmetric in the other (x) direction (with one and two half-waves respectively):

$$\omega_{sa}^2 = \frac{10}{\mu} \frac{0.267\frac{H_x}{l_x^2}\left[63+64\frac{f_y^2}{l_y^2}\right]+\frac{H_y}{l_y^2}\left[4+64\frac{f_x^2}{l_x^2}\right]}{4+64\frac{f_x^2}{l_x^2}+4.06\frac{f_y^2}{l_y^2}}. \tag{6.10}$$

(c) Antisymmetric vibration form (with two half waves in both directions):

$$\omega_{aa}^2 = \frac{42}{\mu} \frac{\frac{H_x}{l_x^2}\left[1+16\frac{f_y^2}{l_y^2}\right]+\frac{H_y}{l_y^2}\left[1+16\frac{f_x^2}{l_x^2}\right]}{1+16\frac{f_x^2}{l_x^2}+16\frac{f_y^2}{l_y^2}}. \tag{6.11}$$

The vibrating mass of the edge ring slows down the vibration of the net, and hence it reduces the above natural frequencies.

Appendix

A.1. A Basic Summary of Matrix Algebra

A.1.1. Notations

Rectangular matrix:

$$\underset{(mn)}{\mathbf{A}} = \begin{bmatrix} a_{11} & a_{12} & \dots & a_{1n} \\ a_{21} & a_{22} & \dots & a_{2n} \\ \dots & \dots & \dots & \dots \\ a_{m1} & a_{m2} & \dots & a_{mn} \end{bmatrix} = [a_{jk}]; \quad \begin{matrix} (j = 1, 2, \dots, m), \\ (k = 1, 2, \dots, n). \end{matrix}$$

The transpose of the rectangular matrix **A** is:

$$\underset{(nm)}{\mathbf{A}^*} = \begin{bmatrix} a_{11} & a_{21} & \dots & a_{m1} \\ a_{12} & a_{22} & \dots & a_{m2} \\ \dots & \dots & \dots & \dots \\ a_{1n} & a_{2n} & \dots & a_{mn} \end{bmatrix}.$$

Matrix **A** is a square matrix if $m=n$

Symmetrical matrix:

$\mathbf{A} = \mathbf{A}^*$

Diagonal matrix:

$$\underset{(n)}{\mathbf{A}} = \begin{bmatrix} a_{11} & 0 & \dots & 0 \\ 0 & a_{22} & \dots & 0 \\ \dots & \dots & \dots & \dots \\ 0 & 0 & \dots & a_{nn} \end{bmatrix} = \langle a_1 \, a_2 \dots a_n \rangle.$$

Unit matrix:

$$\underset{(n)}{\mathbf{E}} = [\delta_{jk}]; \quad \delta_{jk} = \begin{cases} 1, & \text{if } \ j = k \\ 0, & \text{if } \ j \neq k \end{cases}.$$

All elements of the *zero matrix* are zero.
Column matrix of dimension m: (or column vector)

$$\underset{(m)}{\mathbf{a}} = \begin{bmatrix} a_1 \\ a_2 \\ \vdots \\ a_m \end{bmatrix}.$$

Row matrix (or row vector):

$$\underset{(n)}{\mathbf{b}^*} = [b_1 \;\; b_2 \ldots b_n].$$

Unit vectors of dimension n:
first:

$$\underset{(n)}{\mathbf{e}_1^*} = [1 \;\; 0 \;\; 0 \ldots 0],$$

second:

$$\underset{(n)}{\mathbf{e}_2^*} = [0 \;\; 1 \;\; 0 \ldots 0],$$

nth:

$$\underset{(n)}{\mathbf{e}_n^*} = [0 \;\; 0 \;\; 0 \ldots 1].$$

A few *partioning* variations:

$$\mathbf{A} = \begin{bmatrix} a_{11} & a_{12} & a_{13} \\ a_{21} & a_{22} & a_{23} \\ a_{31} & a_{32} & a_{33} \end{bmatrix} = \begin{bmatrix} \mathbf{a}_1^* \\ \mathbf{a}_2^* \\ \mathbf{a}_3^* \end{bmatrix} = \begin{bmatrix} \mathbf{B} & \mathbf{c} \\ \mathbf{d}^* & \mathbf{a}_{33} \end{bmatrix}$$

$$\mathbf{a}_1^* = [a_{11} \;\; a_{12} \;\; a_{13}]; \quad \mathbf{d}^* = [a_{31} \;\; a_{32}]; \quad \mathbf{c} = \begin{bmatrix} a_{13} \\ a_{23} \end{bmatrix};$$

$$\mathbf{B} = \begin{bmatrix} a_{11} & a_{12} \\ a_{21} & a_{22} \end{bmatrix} = \mathbf{A}_{33}.$$

($\mathbf{A}_{33}$ is the minor-matrix belonging to element a_{33} of matrix **A**).

A.1.2. Rules of Arithmetic

Equality:

$$\underset{(mn)}{\mathbf{A}} = \underset{(mn)}{\mathbf{B}}, \quad \text{if} \quad a_{ij} = b_{ij} \begin{pmatrix} i = 1, 2, \ldots, m) \\ j = 1, 2, \ldots, n) \end{pmatrix}.$$

Addition:

$$\mathbf{C} = \mathbf{A}+\mathbf{B} = [a_{jk}]+[b_{jk}] = [c_{jk}],$$

$$c_{jk} = a_{jk}+b_{jk}.$$

Multiplication by a scalar:

$$\alpha\mathbf{A} = \alpha[a_{jk}] = [\alpha a_{jk}].$$

Transpose of a sum:

$$(\mathbf{A}+\mathbf{B})^* = \mathbf{A}^*+\mathbf{B}^*.$$

Matrix multiplication:

$$\mathbf{AB} = \mathbf{C}$$

$$\mathbf{A} = [a_{jk}]; \quad \begin{pmatrix} j = 1, 2, \ldots, m \\ k = 1, 2, \ldots, r \end{pmatrix}$$

$$\mathbf{B} = [b_{ih}]; \quad \begin{pmatrix} i = 1, 2, \ldots, r \\ h = 1, 2, \ldots, n \end{pmatrix}$$

$$\mathbf{C} = [c_{st}]; \quad \begin{pmatrix} s = 1, 2, \ldots, m \\ t = 1, 2, \ldots, n \end{pmatrix}$$

$$c_{st} = \sum_{v=1}^{r} a_{sv} b_{vt}.$$

Dyadic multiplication:

$$\mathbf{ab}^* = \begin{bmatrix} a_1 \\ a_2 \\ \vdots \\ a_m \end{bmatrix} [b_1 \;\; b_2 \ldots b_n] = \begin{bmatrix} a_1 b_1 & a_1 b_2 & \ldots & a_1 b_n \\ a_2 b_1 & a_2 b_2 & \ldots & a_2 b_n \\ \ldots & \ldots & \ldots & \ldots \\ a_m b_1 & a_m b_2 & \ldots & a_m b_n \end{bmatrix}.$$

Scalar product of row and column matrices:

$$\mathbf{b}^*\mathbf{c} = [b_1 \;\; b_2 \ldots b_n] \begin{bmatrix} c_1 \\ c_2 \\ \vdots \\ c_n \end{bmatrix} = \sum_{i=1}^{n} b_i c_i.$$

Transpose of a matrix-product:

$$(\mathbf{AB})^* = \mathbf{B}^*\mathbf{A}^*.$$

The elements of a *hypermatrix* are matrices. E.g.:

$$\mathbf{A} = \begin{bmatrix} \mathbf{A}_{11} & \mathbf{A}_{12} \\ \mathbf{A}_{21} & \mathbf{A}_{22} \end{bmatrix}; \quad \mathbf{B} = \begin{bmatrix} \mathbf{B}_{11} & \mathbf{B}_{12} \\ \mathbf{B}_{21} & \mathbf{B}_{22} \end{bmatrix}$$

their product (if the multiplication of the elements can be carried out):

$$\mathbf{A}\times\mathbf{B} = \mathbf{C} = \begin{bmatrix} \mathbf{C}_{11} & \mathbf{C}_{12} \\ \mathbf{C}_{21} & \mathbf{C}_{22} \end{bmatrix},$$

where, e.g.:

$$\mathbf{C}_{11} = \mathbf{A}_{11}\mathbf{B}_{11}+\mathbf{A}_{12}\mathbf{B}_{21}, \text{ etc.}$$

Minor matrix:
The minor matrix belonging to element a_{jk} of matrix $\mathbf{A}$:

$$\mathbf{A}_{jk} = \begin{bmatrix} a_{11} & a_{12} & \dots & a_{1,k-1} & a_{1,k+1} & \dots & a_{1n} \\ a_{21} & a_{22} & \dots & a_{2,k-1} & a_{2,k+1} & \dots & a_{2n} \\ \dots & \dots & \dots & \dots & \dots & \dots & \dots \\ a_{j-1,1} & a_{j-1,2} & \dots & a_{j-1,k-1} & a_{j-1,k+1} & \dots & a_{j-1,n} \\ a_{j+1,1} & a_{j+1,2} & \dots & a_{j+1,k-1} & a_{j+1,k+1} & \dots & a_{j+1,n} \\ \dots & \dots & \dots & \dots & \dots & \dots & \dots \\ a_{m1} & a_{m2} & \dots & a_{m,k-1} & a_{m,k+1} & \dots & a_{mn} \end{bmatrix}.$$

Determinant of a matrix:

$$\mathbf{A} = \begin{bmatrix} a_{11} & a_{12} & \dots & a_{1n} \\ a_{21} & a_{22} & \dots & a_{2n} \\ \dots & \dots & \dots & \dots \\ a_{n1} & a_{n2} & \dots & a_{nn} \end{bmatrix}$$

is

$$\det \mathbf{A} = |\mathbf{A}| = \begin{vmatrix} a_{11} & a_{12} & \dots & a_{1n} \\ a_{21} & a_{22} & \dots & a_{2n} \\ \dots & \dots & \dots & \dots \\ a_{n1} & a_{n2} & \dots & a_{nn} \end{vmatrix}.$$

Determinant of a matrix-product:

$$|\mathbf{AB}| = |\mathbf{A}|\times|\mathbf{B}|.$$

Adjoint of a square matrix:

$$\text{adj } \mathbf{A} = \begin{bmatrix} |\mathbf{A}_{11}| & |\mathbf{A}_{21}| & \dots & |\mathbf{A}_{n1}| \\ |\mathbf{A}_{12}| & |\mathbf{A}_{22}| & \dots & |\mathbf{A}_{n2}| \\ \dots & \dots & \dots & \dots \\ |\mathbf{A}_{1n}| & |\mathbf{A}_{2n}| & \dots & |\mathbf{A}_{nn}| \end{bmatrix}.$$

Inverse of a square matrix:

The inverse of matrix $\mathbf{A}$ is $\mathbf{A}^{-1}=\mathbf{B}$, if

$$\mathbf{AB} = \mathbf{BA} = \mathbf{E} \quad \text{and} \quad |\mathbf{A}| \neq 0,$$

$$\mathbf{A}^{-1} = \frac{1}{|\mathbf{A}|} \operatorname{adj} \mathbf{A} = \mathbf{B}.$$

The inverse of a matrix-product:

$$(\mathbf{AB})^{-1} = \mathbf{B}^{-1}\mathbf{A}^{-1}.$$

Singular matrix:

$$\mathbf{A} = 0.$$

Projector-matrix:

$$\mathbf{AA} = \mathbf{A}^2 = \mathbf{A}.$$

Orthogonal matrix:

$$\mathbf{AA}^* = \mathbf{A}^*\mathbf{A} = \mathbf{E},$$

$$\mathbf{A}^{-1} = \mathbf{A}^*.$$

Polynomial of a matrix:

$$P_m\underset{(nn)}{(\mathbf{A})} = \alpha_0\mathbf{E}+\alpha_1\mathbf{A}+\alpha_2\mathbf{A}^2+\ldots+\alpha_m\mathbf{A}^m = \sum_{\nu=0}^{m} \alpha_\nu \mathbf{A}^\nu; \quad (\mathbf{A}^0 = \mathbf{E}).$$

Inverse of a hypermatrix:

$$\mathbf{A} = \begin{bmatrix} \mathbf{A}_{11} & \mathbf{A}_{12} \\ \mathbf{A}_{21} & \mathbf{A}_{22} \end{bmatrix},$$

$$\mathbf{A}^{-1} = \mathbf{B} = \begin{bmatrix} \mathbf{B}_{11} & \mathbf{B}_{12} \\ \mathbf{B}_{21} & \mathbf{B}_{22} \end{bmatrix},$$

if

$$\mathbf{AB} = \mathbf{BA} = \mathbf{E} = \begin{bmatrix} \mathbf{E}_{11} & 0 \\ 0 & \mathbf{E}_{22} \end{bmatrix}.$$

The elements of matrix $\mathbf{B}$ can be calculated from the system of equations

$$\mathbf{B}_{11}\mathbf{A}_{11}+\mathbf{B}_{12}\mathbf{A}_{21} = \mathbf{E}_{11},$$

$$\mathbf{B}_{21}\mathbf{A}_{11}+\mathbf{B}_{22}\mathbf{A}_{21} = 0,$$

$$\mathbf{A}_{11}\mathbf{B}_{12}+\mathbf{A}_{12}\mathbf{B}_{22} = 0,$$

$$\mathbf{A}_{21}\mathbf{B}_{12}+\mathbf{A}_{22}\mathbf{B}_{22} = \mathbf{E}_{22}.$$

If for example $|\mathbf{A}_{11}| \neq 0$, then

$$\mathbf{B}_{22} = (\mathbf{A}_{22} - \mathbf{A}_{21}\mathbf{A}_{11}^{-1}\mathbf{A}_{12})^{-1},$$

$$\mathbf{B}_{12} = -\mathbf{A}_{11}^{-1}\mathbf{A}_{12}\mathbf{B}_{22},$$

$$\mathbf{B}_{21} = -\mathbf{B}_{22}\mathbf{A}_{21}\mathbf{A}_{11}^{-1},$$

$$\mathbf{B}_{11} = (\mathbf{E} - \mathbf{B}_{12}\mathbf{A}_{21})\mathbf{A}_{11}^{-1}.$$

Trace of a square matrix:

$$Tr\mathbf{A} = a_{11} + a_{22} + \ldots + a_{nn},$$

where

$$\underset{(nn)}{\mathbf{A}} = [a_{jk}]; \quad (j, k = 1, 2, \ldots, n).$$

Matrix norms

(a) $N(\mathbf{A}) = \sqrt{Tr(\mathbf{A}\mathbf{A}^*)} = \sqrt{\sum_{j=1}^{m} \sum_{k=1}^{n} a_{jk}^2},$

(b) $N_1(\mathbf{A}) = \sum_{j=1}^{m} \sum_{k=1}^{n} |a_{j,k}|.$

The magnitude (length) of the vector:

$$\|\mathbf{a}\| = \sqrt{\mathbf{a}^*\mathbf{a}} = N(\mathbf{a}).$$

The angle of two vectors of identical dimension:

$$\cos(\mathbf{a}, \mathbf{b}) = \frac{\mathbf{a}^*\mathbf{b}}{\|\mathbf{a}\| \times \|\mathbf{b}\|}.$$

The rank of the matrix:

(a) Rank $\varrho(\mathbf{A})$ of matrix $\mathbf{A}$ is the order of the non-singular square minor matrix of the *highest* order that can be selected from matrix $\mathbf{A}$.

(b) Rank $\varrho(\mathbf{A})$ of matrix $\mathbf{A}$ is the *minimum* number of the dyads necessary for the so-called dyadic generation of matrix $\mathbf{A}$ in the form

$$\mathbf{A} = \mathbf{u}_1\mathbf{v}_1^* + \mathbf{u}_2\mathbf{v}_2^* + \ldots + \mathbf{u}_r\mathbf{v}_r^*.$$

Expansion of a matrix into a minimum number of dyads

Let $a_{11} \neq 0$ in the matrix

$$\mathbf{A} = \begin{bmatrix} a_{11} & a_{12} & a_{13} & \ldots & a_{1n} \\ a_{21} & a_{22} & a_{23} & \ldots & a_{2n} \\ a_{31} & a_{32} & a_{33} & \ldots & a_{3n} \\ \ldots & \ldots & \ldots & \ldots & \ldots \\ a_{m1} & a_{m2} & a_{m3} & \ldots & a_{mn} \end{bmatrix}$$

(if not all the elements of matrix **A** are zero, i.e. if **A** is not a zero matrix, then this can always be achieved by a suitable interchange of rows and columns), and let us form the matrix

$$\mathbf{A}-\frac{1}{a_{11}}\begin{bmatrix} a_{11} \\ a_{21} \\ a_{31} \\ \vdots \\ a_{m1} \end{bmatrix} [a_{11}\ \ a_{12}\ \ a_{13} \dots a_{1n}] = \mathbf{B} = \mathbf{A}-\mathbf{u}_1\mathbf{v}_1^*.$$

Let $b_{22} \neq 0$ in the matrix **B** obtained in this way

$$\mathbf{B} = \begin{bmatrix} 0 & 0 & 0 & \dots 0 \\ 0 & b_{22} & b_{23} & \dots b_{2n} \\ 0 & b_{32} & b_{33} & \dots b_{3n} \\ \dots & \dots & \dots & \dots \\ 0 & b_{m2} & b_{m3} & \dots b_{mn} \end{bmatrix}.$$

Let us form the matrix

$$\mathbf{B}-\frac{1}{b_{22}}\begin{bmatrix} 0 \\ b_{22} \\ b_{32} \\ \vdots \\ b_{m2} \end{bmatrix} [0\ \ b_{22}\ \ b_{23} \dots b_{2n}] = \mathbf{C} = \mathbf{B}-\mathbf{u}_2\mathbf{v}_2^*.$$

Let $c_{33} \neq 0$ in the matrix **C** obtained in this way

$$\mathbf{C} = \begin{bmatrix} 0 & 0 & 0 & \dots 0 \\ 0 & 0 & 0 & \dots 0 \\ 0 & 0 & c_{33} & \dots c_{3n} \\ \dots & \dots & \dots & \dots \\ 0 & 0 & c_{m3} & \dots c_{mn} \end{bmatrix}.$$

Let us form the matrix

$$\mathbf{C}-\frac{1}{c_{33}}\begin{bmatrix} 0 \\ 0 \\ c_{33} \\ \vdots \\ c_{m3} \end{bmatrix} [0\ \ 0\ \ c_{33} \dots c_{3n}] = \mathbf{D} = \mathbf{C}-\mathbf{u}_3\mathbf{v}_3^*.$$

The procedure should be continued until ultimately a matrix $\mathbf{R}=\mathbf{0}$ is obtained:

$$\mathbf{R} = \mathbf{Q}-\mathbf{u}_r\mathbf{v}_r^* = 0.$$

s a function of the variables $x_1, x_2, \ldots, x_s$, if

$$a_{jk} = a_{jk}(x_1, x_2, x_3, \ldots, x_s).$$

ts symbol is:

$$\mathbf{A} = \mathbf{A}(x_1, x_2, x_3, \ldots, x_s).$$

The derivative of a matrix is:

$$\frac{\partial}{\partial x_i}\mathbf{A} = \left[\frac{\partial}{\partial x_i}a_{jk}\right].$$

The derivative of a matrix-product is:

$$\frac{\partial}{\partial x_i}\mathbf{AB} = \left(\frac{\partial}{\partial x_i}\mathbf{A}\right)\mathbf{B} + \mathbf{A}\left(\frac{\partial}{\partial x_i}\mathbf{B}\right).$$

The derivative of a vector function of a vector is:

$$\mathbf{r} = \mathbf{r}(\mathbf{x}); \quad \mathbf{r} = \begin{bmatrix} r_1 \\ r_2 \\ \vdots \\ r_m \end{bmatrix}; \quad \mathbf{x} = \begin{bmatrix} x_1 \\ x_2 \\ \vdots \\ x_n \end{bmatrix}; \quad r_i = r_i(x_1, x_2, \ldots, x_n)$$

$$\frac{\partial \mathbf{r}}{\partial \mathbf{x}} = \left[\frac{\partial}{\partial x_1}\mathbf{r} \ \frac{\partial}{\partial x_2}\mathbf{r} \ \frac{\partial}{\partial x_3}\mathbf{r} \ldots \frac{\partial}{\partial x_n}\mathbf{r}\right] =$$

$$= \begin{bmatrix} \frac{\partial r_1}{\partial x_1} & \frac{\partial r_1}{\partial x_2} & \frac{\partial r_1}{\partial x_3} & \cdots & \frac{\partial r_1}{\partial x_n} \\ \frac{\partial r_2}{\partial x_1} & \frac{\partial r_2}{\partial x_2} & \frac{\partial r_2}{\partial x_3} & \cdots & \frac{\partial r_2}{\partial x_n} \\ \cdots & \cdots & \cdots & \cdots & \cdots \\ \frac{\partial r_m}{\partial x_1} & \frac{\partial r_m}{\partial x_2} & \frac{\partial r_m}{\partial x_3} & \cdots & \frac{\partial r_m}{\partial x_n} \end{bmatrix} = \left[\frac{\partial r_j}{\partial x_k}\right]; \quad \begin{pmatrix} j = 1, 2, \ldots, m \\ k = 1, 2, \ldots, n \end{pmatrix}.$$

The differential of a vector is:

$$d\mathbf{r}(\mathbf{x}) = \sum_{k=1}^{n} \frac{\partial \mathbf{r}}{\partial x_k} dx_k = \frac{\partial \mathbf{r}}{\partial \mathbf{x}} d\mathbf{x}.$$

The definite integral of a matrix function of a scalar parameter
If

$$\mathbf{A} = \mathbf{A}(\lambda) = [a_{jk}(\lambda)] \quad \begin{pmatrix} j = 1, 2, \ldots, m \\ k = 1, 2, \ldots, n \end{pmatrix}$$

The sum of dyads $\mathbf{u}_i\mathbf{v}_i^*$ $(i=1, 2, \ldots, r)$ gives matrix $\mathbf{A}$

$$\mathbf{A} = \sum_{i=1}^{r} \mathbf{u}_i\mathbf{v}^* = [\mathbf{u}_1 \ \mathbf{u}_2 \ \mathbf{u}_3 \ldots \mathbf{u}_r] \begin{bmatrix} \mathbf{v}_1^* \\ \mathbf{v}_2^* \\ \mathbf{v}_3^* \\ \vdots \\ \mathbf{v}_r^* \end{bmatrix} = \mathbf{UV}$$

the rank of which equals r

$$\varrho(\mathbf{A}) = r.$$

In general, $\mathbf{U}$ and $\mathbf{V}$ are lower and upper trapezoidal matrices, the elements of which, u_{ii}, and v_{ii}, respectively $(i=1, 2, \ldots, r)$ elements

$$\mathbf{U} = \begin{bmatrix} u_{11} & 0 & 0 & \ldots 0 \\ u_{12} & u_{22} & 0 & \ldots 0 \\ u_{13} & u_{23} & u_{33} & \ldots 0 \\ \vdots & \vdots & \vdots & \vdots \\ u_{1r} & u_{2r} & u_{3r} & \ldots u_{rr} \\ \vdots & \vdots & \vdots & \vdots \\ u_{1m} & u_{2m} & u_{3m} & \ldots u_{rm} \end{bmatrix},$$

$$\mathbf{V} = \begin{bmatrix} v_{11} & v_{12} & v_{13} \ldots v_{1r} \ldots v_{1n} \\ 0 & v_{22} & v_{23} \ldots v_{2r} \ldots v_{2n} \\ 0 & 0 & v_{33} \ldots v_{3r} \ldots v_{3n} \\ \ldots & \ldots & \ldots \\ 0 & 0 & 0 \ \ldots v_{rr} \ldots v_{rn} \end{bmatrix}.$$

$\mathbf{U}$ and $\mathbf{V}$ used in the dyadic expansion of the non-singular sq $\underset{(nn)}{\mathbf{A}}$ are lower and upper triangular matrices, respectively.

The inverse of a non-singular square matrix generated in dyadic

$$\mathbf{A} = \mathbf{UV}; \quad |\mathbf{A}| \neq 0,$$

$$\mathbf{A}^{-1} = \mathbf{V}^{-1}\mathbf{U}^{-1}.$$

A.1.3. The Elements of Matrix Analysis

Matrix

$$\mathbf{A} = [a_{jk}] \quad \begin{pmatrix} j = 1, 2, \ldots, m \\ k = 1, 2, \ldots, n \end{pmatrix}$$

then

$$\int_{\lambda_1}^{\lambda_2} \mathbf{A}(\lambda)\,\mathrm{d}\lambda = \left[\int_{\lambda_1}^{\lambda_2} a_{jk}(\lambda)\,\mathrm{d}\lambda\right].$$

Solution of systems of simultaneous linear equations
Equation

$$\mathbf{Ax}+\mathbf{b} = 0$$

in which

$$\mathbf{A} = [a_{j,k}] \quad \begin{pmatrix} j = 1, 2, \ldots, m \\ k = 1, 2, \ldots, n \end{pmatrix} \quad \text{is}$$

$$\left.\begin{array}{l} \text{1. determinate} \\ \text{2. overdeterminate,} \\ \text{3. indeterminate} \end{array}\right\} \quad \text{if} \quad \varrho(\mathbf{A}) = \begin{cases} m = n \\ n < m \\ m < n \end{cases}$$

(If $\varrho(\mathbf{A}) < m, n$, then the equation is partly indeterminate, partly overdeterminate.)
In case 1, the solution is

$$\mathbf{x} = -\mathbf{A}^{-1}\mathbf{b}.$$

In case 2, the equation can be transformed (by suitable rearrangement) into the form

$$\begin{bmatrix}\mathbf{A}_1 \\ \mathbf{A}_2\end{bmatrix}\mathbf{x}+\begin{bmatrix}\mathbf{b}_1 \\ \mathbf{b}_2\end{bmatrix} = 0 \quad (|\mathbf{A}_1| \neq 0),$$

where the solution

$$\mathbf{x} = -\mathbf{A}^{-1}\mathbf{b}_1$$

satisfies the equation consistently only if

$$\mathbf{A}_2\mathbf{x}+\mathbf{b}_2 = 0,$$

i.e.

$$\mathbf{b}_2 = \mathbf{A}_2\mathbf{A}_1^{-1}\mathbf{b}_1.$$

Otherwise the equation can be satisfied only with error.
In case 3, the equation can be transformed (by suitable rearrangement) into the form

$$[\mathbf{A}_{(1)} \;\; \mathbf{A}_{(2)}]\begin{bmatrix}\mathbf{x}_{(1)} \\ \mathbf{x}_{(2)}\end{bmatrix}+\mathbf{b} = 0 \quad (|\mathbf{A}_{(1)}| \neq 0),$$

where

$$\mathbf{x}_{(1)} = -\mathbf{A}_{(1)}^{-1}(\mathbf{A}_{(2)}\mathbf{x}_{(2)}+\mathbf{b}),$$

and $\mathbf{x}_{(2)}$ can be selected arbitrarily.

Solution of a homogeneous system of equations:

$$\mathbf{Ax} = 0$$

has always the so-called trivial solution

$$x_1 = x_2 = \dots = x_n = 0.$$

And if $|\mathbf{A}|=0$ and $\varrho(A)\neq 0$, then an infinite number of solutions exists besides this. In this case, the system of equations — after a possible rearrangement of rows and columns — can always be written in the form

$$\begin{bmatrix} \mathbf{A}_{11} & \mathbf{A}_{12} \\ \mathbf{A}_{21} & \mathbf{A}_{22} \end{bmatrix} \begin{bmatrix} \mathbf{x}_1 \\ \mathbf{x}_2 \end{bmatrix} = 0,$$

where $|\mathbf{A}_{11}|\neq 0$ and $\varrho(\mathbf{A}_{11})=\varrho(\mathbf{A})$. For $\mathbf{x}_2$ arbitrarily chosen on the basis of the above equation it is possible to determine the solution

$$\mathbf{x}_1 = -\mathbf{A}_{11}^{-1}\mathbf{A}_{12}\mathbf{x}_2.$$

Of course, vectors $\mathbf{x}_1$ and $\mathbf{x}_2$ calculated in this way and arbitrarily chosen, satisfy equation

$$\mathbf{A}_{21}\mathbf{x}_1+\mathbf{A}_{22}\mathbf{x}_2 = 0$$

as well, because when writing the above expression of $\mathbf{x}_1$ into the equation, it is easy to see that the relationship

$$\mathbf{A}_{22} = \mathbf{A}_{21}\mathbf{A}_{11}^{-1}\mathbf{A}_{12}$$

must be valid for the right-hand lower submatrix (block) of matrix $\mathbf{A}$. Nevertheless, this is the direct consequence of the requirement $\varrho(\mathbf{A}_{11})= =\varrho(\mathbf{A})$:

$$\mathbf{A}_{21}\mathbf{A}_{11}^{-1}[\mathbf{A}_{11} \;\; \mathbf{A}_{12}] = [\mathbf{A}_{21} \;\; \mathbf{A}_{21}\mathbf{A}_{11}^{-1}\mathbf{A}_{12}] = [\mathbf{A}_{21} \;\; \mathbf{A}_{22}],$$

i.e. the last row $n-\varrho$ of $\mathbf{A}$ is the linear combination of the first row ϱ.

Solution of an overdeterminate system of equations on the basis of the principle of minimum mean-square error

The solution of the overdeterminate equation

$$\mathbf{Ax}+\mathbf{b} = \mathbf{0}; \quad \mathbf{A} = [a_{j,k}]; \quad \mathbf{b} = [b_j]; \quad \mathbf{x} = [x_k] \quad \begin{pmatrix} j = 1, 2, \dots, m \\ k = 1, 2, \dots, n \end{pmatrix}$$

satisfies the equation with error $\mathbf{y}$, i.e. in general

$$\mathbf{Ax}+\mathbf{b} = \mathbf{y}.$$

From among the possible solutions, the one which makes the scalar product $\mathbf{y}^*\mathbf{y}$ of vector $\mathbf{y}$ formed by itself (i.e. the square sum of its elements) minimum should be selected:

$$f(\mathbf{x}) = \mathbf{y}^*\mathbf{y} = \text{min!}$$

Function $f(\mathbf{x})$ can have an extreme value when

$$\frac{\partial}{\partial x_k} f(\mathbf{x}) = 0 \quad (k = 1, 2, \ldots, n),$$

i.e.

$$\frac{\partial}{\partial x_k}(\mathbf{x}^*\mathbf{A}^*+\mathbf{b}^*)(\mathbf{Ax}+\mathbf{b}) = \frac{\partial}{\partial x_k}(\mathbf{x}^*\mathbf{A}^*\mathbf{Ax}+\mathbf{b}^*\mathbf{Ax}+\mathbf{x}^*\mathbf{A}^*\mathbf{b}+\mathbf{b}^*\mathbf{b}) =$$

$$= \mathbf{e}_k^*\mathbf{A}^*\mathbf{Ax}+\mathbf{x}^*\mathbf{A}^*\mathbf{Ae}_k+\mathbf{b}^*\mathbf{Ae}_k+\mathbf{e}_k^*\mathbf{A}^*\mathbf{b} = 0.$$

And because

$$\mathbf{e}_k^*\mathbf{A}^*\mathbf{Ax} = \mathbf{x}^*\mathbf{A}^*\mathbf{Ae}_k,$$

$$\mathbf{b}^*\mathbf{Ae}_k = \mathbf{e}_k^*\mathbf{A}^*\mathbf{b}.$$

we have

$$\mathbf{e}_k^*(\mathbf{A}^*\mathbf{Ax}+\mathbf{A}^*\mathbf{b}) = 0; \quad (k = 1, 2, \ldots, n).$$

In other words, $f(\mathbf{x})$ can have an extreme value when

$$\mathbf{A}^*\mathbf{Ax}+\mathbf{A}^*\mathbf{b} = 0.$$

Thus,

$$\mathbf{x} = -(\mathbf{A}^*\mathbf{A})^{-1}\mathbf{A}^*\mathbf{b}$$

is that solution of the overdeterminate equation which satisfies the equation with minimum mean-square error.

Eigenvalues and eigenvectors of a square matrix:

$$\mathbf{Au}_j = \lambda_j\mathbf{u}_j \quad (j, k = 1, 2, \ldots, n).$$

$$\mathbf{v}_k^*\mathbf{A} = \lambda_k\mathbf{v}_k^*$$

λ_j, λ_k, are the jth and kth eigenvalues, respectively, of matrix $\mathbf{A}$, which can be calculated from the characteristic equation

$$|\mathbf{A}-\lambda_i\mathbf{E}| = 0 \quad (i = 1, 2, \ldots, n).$$

The jth right vectors $\mathbf{u}_j$ and the kth left vectors $\mathbf{v}_k^*$, respectively, can be calculated from the homogeneous system of equations

$$(\mathbf{A}-\lambda_j\mathbf{E})\mathbf{u}_j = 0,$$

and

$$(\mathbf{A}^*-\lambda_k\mathbf{E})\mathbf{v}_k = 0,$$

respectively. The right and left vectors are orthogonal:

$$\mathbf{v}_k^*\mathbf{u}_j = 0, \quad \text{if} \quad j \neq k.$$

A.2. Equilibrium and Displacement Equations of a Bar with Spatially Curved Axis

The axis of the bar i shown in Fig. A.1 is a space curve. The origin of the bar is at joint j with position vector $\mathbf{r}_j$, while its end point is at joint k with position vector $\mathbf{r}_k$. The length of the bar measured along its axis is l_i. The points on the bar axis are given by the position vector $\mathbf{r}_\lambda$, where λ is the distance of the point from the origin, measured along the axis, so that $0 \leq \lambda \leq l_i$.

$$\mathbf{r}_j = \begin{bmatrix} x_j \\ y_j \\ z_j \end{bmatrix}; \quad \mathbf{r}_k = \begin{bmatrix} x_k \\ y_k \\ z_k \end{bmatrix}; \quad \mathbf{r}_\lambda = \begin{bmatrix} x_\lambda \\ y_\lambda \\ z_\lambda \end{bmatrix}.$$

The bar cross-section corresponding to the axis point with parameter λ is a plane section perpendicular to the tangent of the bar axis; the normal and

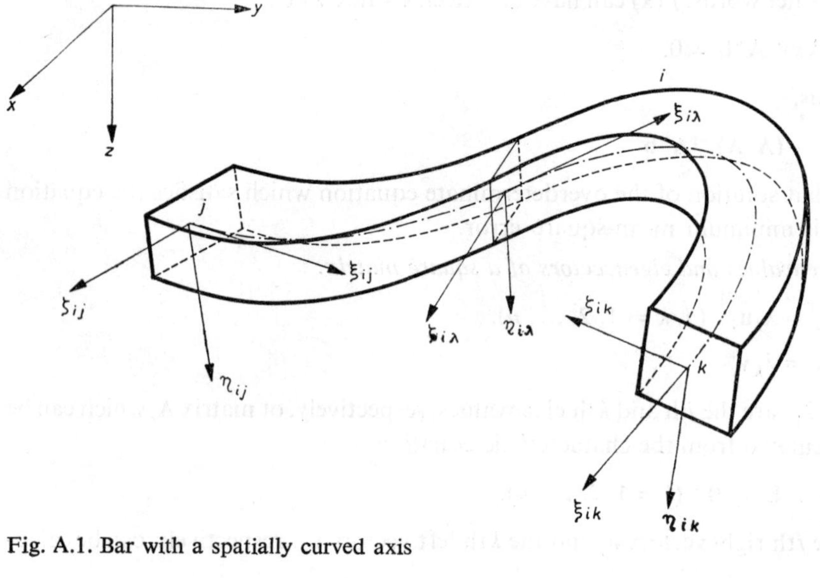

Fig. A.1. Bar with a spatially curved axis

the two principal axes of inertia are the three axes $\xi_{i\lambda}$, $\eta_{i\lambda}$, $\zeta_{i\lambda}$ allocated to the axis point (see Fig. A.1).
The components of the internal force and couple resultants, belonging to the cross-section at position λ, can be summarized in a single vector $\mathbf{s}_{i\lambda}$:

$$\mathbf{s}_{i\lambda} = \begin{bmatrix} P_{\xi i\lambda} \\ P_{\eta i\lambda} \\ P_{\zeta i\lambda} \\ M_{\xi i\lambda} \\ M_{\eta i\lambda} \\ M_{\zeta i\lambda} \end{bmatrix}.$$

Due to $\mathbf{s}_{i\lambda}$, a deformation given by the product

$$\mathbf{f}_{i\lambda}\mathbf{s}_{i\lambda}\,d\lambda$$

occurs along the prism of height $d\lambda$ corresponding to the cross-section at position λ, where $\mathbf{f}_{i\lambda}$ is the diagonal matrix of specific elasticity allocated to the cross-section at position λ:

$$\mathbf{f}_{i\lambda} = \left\langle \frac{1}{EA_{i\lambda}}\,\frac{\varrho_{\eta i\lambda}}{GA_{i\lambda}}\,\frac{\varrho_{\zeta i\lambda}}{GA_{i\lambda}}\,\frac{1}{GJ_{\xi i\lambda}}\,\frac{1}{EJ_{\eta i\lambda}}\,\frac{1}{EJ_{\zeta i\lambda}} \right\rangle,$$

in which

- E, G are the moduli of elasticity and shear, respectively;
- $A_{i\lambda}$ is the area of the cross-section characterized by λ of bar i;
- $J_{\eta i\lambda}, J_{\xi i\lambda}$ are the moments of inertia referred to the axes η and ζ, respectively, passing through the centre of gravity of the cross-section at position λ;
- $\varrho_{\eta i\lambda}, \varrho_{\zeta i\lambda}$ are the shape factors of the cross-section at position λ, corresponding to the η- and ζ-directional shear forces, respectively;
- $GJ_{\xi i\lambda}$ is the pure torsional rigidity of the bar cross-section at position λ.

The bar is acted upon at its axis point λ by a concentrated external force vector

$$\mathbf{q}_{\lambda} = \begin{bmatrix} R_{x\lambda} \\ R_{y\lambda} \\ R_{z\lambda} \\ N_{x\lambda} \\ N_{y\lambda} \\ N_{z\lambda} \end{bmatrix} \begin{matrix} \left.\begin{matrix} \\ \\ \\ \end{matrix}\right\} \text{force components} \\ \left.\begin{matrix} \\ \\ \\ \end{matrix}\right\} \text{couple components} \end{matrix}$$

or by a force, continuously distributed along the axis of the bar, of intensity given by the vector

$$\mathbf{g}_\lambda = \begin{bmatrix} g_{x\lambda} \\ g_{y\lambda} \\ g_{z\lambda} \\ \gamma_{x\lambda} \\ \gamma_{y\lambda} \\ \gamma_{z\lambda} \end{bmatrix} \begin{matrix} \left.\begin{matrix} \\ \\ \\ \end{matrix}\right\} \text{distributed force components} \\ \left.\begin{matrix} \\ \\ \\ \end{matrix}\right\} \text{distributed couple components.} \end{matrix}$$

Hereinafter, the concentrated external force $\mathbf{q}_\lambda$ allocated to parameter λ will be given in the form $\mathbf{q}=\mathbf{g}_\lambda\, d\lambda$.

The external force acting at the end point k of the bar equals the value of the internal force $\mathbf{s}_i$ allocated to the parameter $\lambda=l_i$. Force

$$\mathbf{s}_i = \mathbf{s}_{i\lambda}\big|_{\lambda=l_i} = \mathbf{s}_{ik}$$

is termed bar force $\mathbf{s}_i$ of bar i. Bar force $\mathbf{s}_i$ and $\mathbf{q}_\lambda=\mathbf{g}_\lambda\, d\lambda$ together determine the internal force $\mathbf{s}_{i\lambda_1}$ (internal force of the cross-section) corresponding to any axis point of bar i in the position λ_1:

$$\mathbf{s}_{i\lambda_1} = \mathbf{T}_{i\lambda_1}\mathbf{B}^*_{\lambda_1 k}\mathbf{T}^*_{ik}\mathbf{s}_i + \int_{\lambda_1}^{l_i} \mathbf{T}_{i\lambda_1}\mathbf{B}^*_{\lambda_1\lambda}\mathbf{g}_\lambda\, d\lambda, \tag{A.2.1}$$

where $\mathbf{T}$ depends only on the position of the axis cross-section corresponding to the axis point, and $\mathbf{B}$ on the position vectors of the axis points:

$$\mathbf{T}_{i\lambda} = \begin{bmatrix} \mathbf{T}_{0;i\lambda} & \\ & \mathbf{T}_{0;i\lambda} \end{bmatrix}; \quad \mathbf{T}_{0;i\lambda} = \begin{bmatrix} \mathbf{e}^*_{\xi i\lambda} \\ \mathbf{e}^*_{\eta i\lambda} \\ \mathbf{e}^*_{\zeta i\lambda} \end{bmatrix};$$

$$\mathbf{B}_{\lambda_1\lambda} = \begin{bmatrix} \mathbf{E} & \mathbf{B}_{0;\lambda_1\lambda} \\ 0 & \mathbf{E} \end{bmatrix};$$

$$\mathbf{B}_{0;\lambda_1\lambda} = \begin{bmatrix} 0 & z_\lambda - z_{\lambda_1} & y_{\lambda_1} - y_\lambda \\ z_{\lambda_1} - z_\lambda & 0 & x_\lambda - x_{\lambda_1} \\ y_\lambda - y_{\lambda_1} & x_{\lambda_1} - x_\lambda & 0 \end{bmatrix}.$$

The equilibrium equation corresponding to each joint of the bar structure expresses the zero-valency of the (internal) forces transferred to the joint and of the external forces directly acting on the joint. Bar i exerts a force of magnitude

$$-\mathbf{T}^*_{ik}\mathbf{s}_i \tag{A.2.2}$$

expressed in the x, y, z coordinate system, on joint k, and a force

$$\mathbf{B}_{jk}^{*}\mathbf{T}_{ik}^{*}\mathbf{s}_i+\int_0^{l_i}\mathbf{B}_{j\lambda}^{*}\mathbf{g}_{\lambda}\,d\lambda \tag{A.2.3}$$

on joint j. If it is temporarily assumed that only bar i is connected to joints j and k, which are directly acted upon by external forces $\mathbf{q}_i$ and $\mathbf{q}_k$, then the equilibrium equation of the joints can be written in the form

$$\begin{bmatrix}\mathbf{B}_{jk}^{*}\mathbf{T}_{ik}^{*}\\-\mathbf{T}_{ik}^{*}\end{bmatrix}\mathbf{s}_i+\begin{bmatrix}\mathbf{q}_j+\int_0^{l_i}\mathbf{B}_{j\lambda}^{*}\mathbf{g}_{\lambda}\,d\lambda\\\mathbf{q}_k\end{bmatrix}=0. \tag{A.2.4}$$

This shows that the direct load acting on each bar of the structure can be reduced to the joint corresponding to the origin of the bar by expression (A.2.4).

The relationship between the displacement vector of the starting point j of the bar $\mathbf{u}_j$ and that of its end point $\mathbf{u}_k$, given in the x, y, z coordinate system, the elastic deformation of the bar and the non-elastic deformation ($\mathbf{t}$) prescribed for the bar are described by the equation

$$\mathbf{u}_k=\mathbf{B}_{jk}\mathbf{u}_j+\int_0^{l_i}\mathbf{B}_{\lambda_1 k}\mathbf{T}_{i\lambda_1}^{*}\mathbf{f}_{i\lambda_1}\mathbf{s}_{i\lambda_1}\,d\lambda_1+\mathbf{T}_{ik}^{*}\mathbf{t}_{0;i} \tag{A.2.5}$$

where

$$\mathbf{t}_{0;i}=\begin{bmatrix}t_{\xi ik}\\t_{\eta ik}\\t_{\zeta ik}\\\vartheta_{\xi ik}\\\vartheta_{\eta ik}\\\vartheta_{\zeta ik}\end{bmatrix}\begin{matrix}\left.\begin{matrix}\\\\\\\end{matrix}\right\}\text{displacement components}\\\left.\begin{matrix}\\\\\\\end{matrix}\right\}\text{axis cross-displacement components}\end{matrix}$$

is the kinematic load vector allocated to bar i (the non-elastic deformation prescribed for the bar), given in the ξ, η, ζ coordinate system of point k and corresponding to point k. Equation (A.2.5), taking (A.2.1) into consideration and multiplying every term of the equation by $\mathbf{T}_{ik}$, can be written also in the form

$$[\mathbf{T}_{ik}\mathbf{B}_{jk}\quad-\mathbf{T}_{ik}]\begin{bmatrix}\mathbf{u}_j\\\mathbf{u}_k\end{bmatrix}+\mathbf{F}_i\mathbf{s}_i+\mathbf{t}_i=0. \tag{A.2.6}$$

Equation (A.2.6) is called the displacement equation of bar i, in which

$$\mathbf{F}_i=\mathbf{T}_{ik}\left(\int_0^{l_i}\mathbf{B}_{\lambda_1 k}\mathbf{T}_{i\lambda_1}^{*}\mathbf{f}_{i\lambda_1}\mathbf{T}_{i\lambda_1}\mathbf{B}_{\lambda_1 k}^{*}\,d\lambda_1\right)\mathbf{T}_{ik}^{*} \tag{A.2.7}$$

is the flexibility matrix of the bar with space-curve axis and $\mathbf{t}_i$ is the reduced kinematic load vector of the bar

$$\mathbf{t}_i = \mathbf{t}_{0;i} + \mathbf{T}_{ik} \int_0^{l_i} \mathbf{B}_{\lambda_1 k} \mathbf{T}^*_{i\lambda_1} \mathbf{f}_{i\lambda_1} \mathbf{T}_{i\lambda_1} \int_0^{l_i} \mathbf{B}^*_{\lambda_1 \lambda} \mathbf{g}_\lambda \, d\lambda \, d\lambda_1 . \tag{A.2.8}$$

Equation (A.2.6) expresses the relative displacements between the starting and end points of the bar, j, k. The effect of the loading forces acting directly on the bar can be included in the kinematic load vector $\mathbf{t}_i$, with the aid of expression (A.2.8).

It can easily be checked that in the case of a bar having a straight axis and a constant cross-section, integral (A.2.7) is reduced to the following simple expression if the shear deformations are neglected:

$$\mathbf{F}_i = \begin{bmatrix} \frac{l_i}{EA_i} & & & & & \\ & \frac{l_i^3}{3EI_{\zeta i}} & & & & \frac{l_i^2}{2EI_{\zeta i}} \\ & & \frac{l_i^3}{3EI_{\eta i}} & & -\frac{l_i^2}{2EI_{\eta i}} & \\ & & & \frac{l_i}{GI_{\xi i}} & & \\ & & -\frac{l_i^2}{2EI_{\eta i}} & & \frac{l_i}{EI_{\eta i}} & \\ & \frac{l_i^2}{2EI_{\zeta i}} & & & & \frac{l_i}{EI_{\zeta i}} \end{bmatrix} .$$

References

[1] Argyris, J. H. and Angelopoulos, T.: Theorie, Programmentwicklung und Erfahrung an vorgespannten Netzwerkkonstruktionen. IVBH IX. Kongr. Vorbericht, 1972. 377–384.
Ein Verfahren für die Formfindung beliebiger, vorgespannten Netzwerkkonstruktionen. IVBH IX. Kongr. Vorbericht, 1972. 385–392.

[2] Bandel, H. K.: Das orthogonale Seilnetz hyperbolisch–parabolisher Form unter vertikalen Lastzuständen und Temperaturänderung. Der Bauingenieur, **34** (1959), 394–401.

[3] Beutler, J.: Beitrag zur statistischen Windbelastung von Seilnetzwerken—Ergebnisse von Windkanaluntersuchungen. Proceedings of the IASS Colloquium on Hanging Roofs, Continuous Metallic Shell Roofs and Superficial Lattice Roofs, Paris, 1962. North-Holland Publishing Co., Amsterdam. 1962. 78–86.

[4] Bronstein, I. N. and Semedjajev, K.A.: Taschenbuch der Mathematik. G. B. Teubner, Leipzig, 1959.

[5] Dischinger, F.: Elastische und plastische Verformungen der Eisenbetontragwerke und insbesondere der Bogenbrücken. Der Bauingenieur, **20** (1939). 286–294, 426–437, 563–572.

[6] Eras, G. and Elze, H.: Berechnungsverfahren für vorgespannte doppelt gekrümmte Seilnetzwerke. Bauplanung–Bautechnik, **15** (1961), 349–354.

[7] Eras, G. and Elze, H.: Zur Berechnung und statisch vorteilhaften Formgebung von Seilnetzwerken. Proceedings of the IASS Colloquium on Hanging Roofs, Continuous Metallic Shell Roofs and Superficial Lattice Roofs, Paris, 1962. North-Holland Publishing Co., Amsterdam. 1962. 68–75.

[8] Halász, O., Roller, B. and Vértes, Gy.: Függesztett tetőszerkezetek néhány számítási kérdéséről. (On some problems of the calculation of suspended roof structures). Építőipari és Közlekedési Műszaki Egyetem Tudományos Közleményei, Vol. VI, No. 4, Budapest, 1960. 81–111.
Függesztett hálók számítása az elsőrendű elmélet alapján. (Calculation of suspended nets on the basis of the first-order theory.) Mélyépítéstudományi Szemle, **10** (1960), 131–137.

[9] Kollár, L. and Köröndi, L.: Függőtetők tervezése és gyakorlati számítása. (The design and calculation of suspended roofs). Mérnöki Továbbképző Intézet (Postgraduate Training Institute for Engineers), Budapest, 1966.

[9a] Köröndi, L.: Zárt peremgyűrűs függőtető optimális alakjának közelítő meghatározása. (Approximate determination of the optimal shape of suspended roofs having closed edge ring). Dissertation. Budapest, 1978.

[10] Leonhardt, F.: Seilkonstruktionen und seilverspannte Konstruktionen, IVBH IX. Kongr. Einführungsbericht, 1972. 103–125.

[11] Møllmann, H. and Lundhus Mortensen, P.: The Analysis of Prestressed Suspended Roofs (Cable Nets). Space Structures (R. N. Davies, Editor), Blackwell Scientific Publications, Oxford and Edinburgh, 1967.

[12] Otto, F.: Das hängende Dach. Ullstein-Verlag, Berlin, 1954.

[13] Roark, R. J.: Formulas for Stress and Strain. McGraw-Hill, New York, 1965.

[14] Roller, B.: Feszített függőtetők gyakorlati számítása. (Practical calculation of pretensioned suspended roofs). Mélyépítéstudományi Szemle, **15** (1965), 60–69.

[15] Roller, B.: Berechnung von Seilträgernetzen zylindrischer Gestalt, versteift durch Querbalken. Acta Techn. Acad. Sci. Hung. **53** (1966), 407–415.

[16] Schlaich, J., Altmann, H., Bergermann, R., Gabriel, K., Horstkötter, K., Kleinhanss, K., Linhart, P., Mayr, G., Noesgen, J., Otto, U. and Schmidt H.: Das Olympiadach in München. IVBH IX. Kongr. Vorbericht, 1972. 365–376.

[17] Schleyer, F. K.: Über die Berechnung von Seilnetzen. Dissertation. Technische Universität, Berlin, 1960.
Die Berechnung von Seilnetzen. Proceedings of the IASS Colloquium on Hanging Roofs, Continuous Metallic Shell Roofs and Superficial Lattice Roofs, Paris, 1962. North-Holland Publishing Co., Amsterdam. 1962. 48–55.
Die Berechnung von Seilwerken. Proceedings of the IASS Colloquium on Hanging Roofs, Continuous Metallic Shell Roofs and Superficial Lattice Roofs, Paris, 1962. North-Holland Publishing Co., Amsterdam. 1962. 56–61.

[18] Siev, A.: A General Analysis of Prestressed Nets. IVBH Abhandlungen, **23**, (1963), 283–292.

[19] Stahlbau. Ein Handbuch für Studium und Praxis. Bd. II. Deutscher Stahlbau-Verband, Köln, 1964.

[20] Stüssi, F.: Vorlesungen über Baustatik, II. Bd. Statisch unbestimmte Systeme. Birkhäuser, Basel–Stuttgart, 1954.

[21] Szabó, J.: Mit Hilfe der kanonischer Form der Matrixfunktionen vorteilhaft zu behandelnde Aufgaben auf dem Gebiet der Statik und Festigkeitslehre, Wiss. Zeitschr. d. Techn. Univ. Dresden, **10** (1961), 1325–1327.

[22] Szabó, J.: A térbeli tartórács egyenlete. (The equation of spatial girder grillages). ÉTI Tudományos Közlemények, No. 34, Budapest, 1964.

[23] Szabó, J.: Kötélháló állapotterének elemzése. (Analysis of the state space of a cable net). Építés- és Közlekedéstudományi Közlemények, **1** (1965) 315–323.

[24] Szabó, J. and Béres, E.: Beitrag zur Abhandlung über die "Berechnung von Seilträgernetzen" von B. Roller, Acta Techn. Acad. Sci. Hung., **55** (1966), 109–116.

[25] Szabó, J. and Roller, B.: Anwendung der Matrizenrechnung auf Stabwerke. Akadémiai Kiadó, Budapest, 1971.

[26] Szabó, J. and Berényi, M.: Numerical analysis of rectangular cable nets. Acta Techn. Acad. Sci. Hung., **72** (1972), 257–271.

[27] Szabó, J.: Ein Verfahren für die Formfindung vorgespannter Netzwerkkonstruktionen, IVBH IX. Kongr. Schlussbericht, 1972, 139–143.

[28] Szabó, J. and Berényi, M.: Theorie und Praxis der Berechnung von Seilkonstruktionen, IVBH Abhandlungen **33**/II. (1973), 205–220.

[29] Szabó, J.: Bemerkungen zur Berechnung seilverspannter Konstruktionen. Acta Techn. Acad. Sci. Hung., **75** (1973), 357–370.

[30] Szmodits, K.: Függesztett tetőszerkezetek méretezési eljárásai. (Dimensioning procedures of suspended roof structures). ÉTI Tudományos Közlemények, No. 14., Budapest, 1959.

Subject index

DAT